Popular Mechanics

GADGET PLANET

HEARST BOOKS
New York

An Imprint of Sterling Publishing
387 Park Avenue South
New York, NY 10016

Popular Mechanics is a registered trademark of Hearst Communications, Inc.

Book Design: Nancy Leonard

Every effort has been made to ensure that all the information in this book is
accurate. However, due to differing conditions, tools, and individual skills, the
publisher cannot be responsible for any injuries, losses, and/or other damages that
may result from the use of the information in this book.

ISBN 978-1-61837-079-2

Distributed in Canada by Sterling Publishing
c/o Canadian Manda Group, 165 Dufferin Street
Toronto, Ontario, Canada M6K 3H6

Distributed in the United Kingdom by GMC Distribution Services
Castle Place, 166 High Street, Lewes, East Sussex, England BN7 1XU

Distributed in Australia by Capricorn Link (Australia) Pty. Ltd.
P.O. Box 704, Windsor, NSW 2756, Australia

For information about custom editions, special sales,
and premium and corporate purchases, please contact Sterling Special Sales
at 800-805-5489 or specialsales@sterlingpublishing.com.

Manufactured in China

2 4 6 8 10 9 7 5 3 1

www.sterlingpublishing.com

PopularMechanics

GADGET PLANET

150 GIZMOS & INVENTIONS
THAT CHANGED THE WORLD

HEARST BOOKS

New York

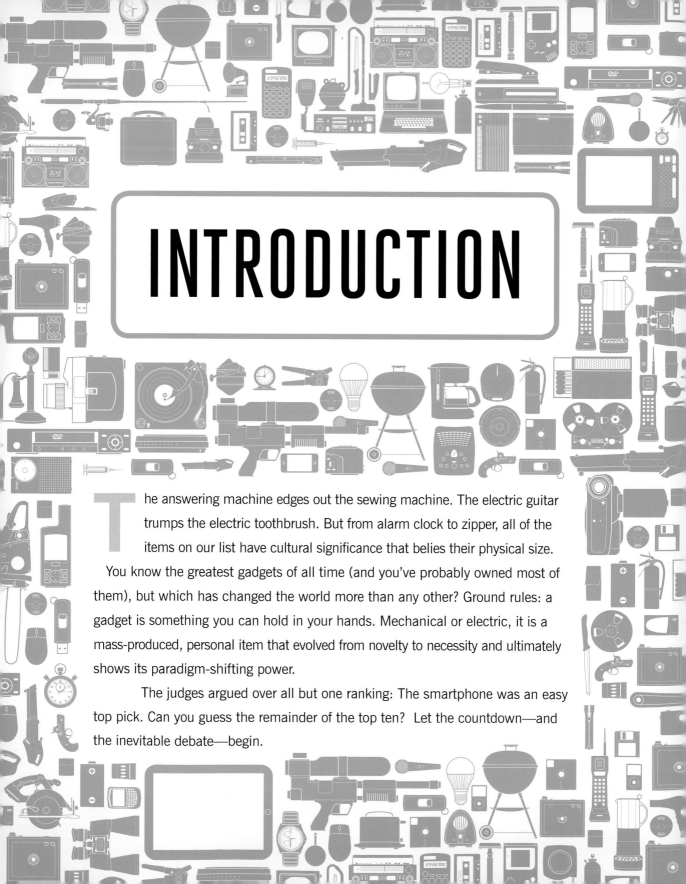

INTRODUCTION

The answering machine edges out the sewing machine. The electric guitar trumps the electric toothbrush. But from alarm clock to zipper, all of the items on our list have cultural significance that belies their physical size. You know the greatest gadgets of all time (and you've probably owned most of them), but which has changed the world more than any other? Ground rules: a gadget is something you can hold in your hands. Mechanical or electric, it is a mass-produced, personal item that evolved from novelty to necessity and ultimately shows its paradigm-shifting power.

The judges argued over all but one ranking: The smartphone was an easy top pick. Can you guess the remainder of the top ten? Let the countdown—and the inevitable debate—begin.

Drawing from the fields of design, technology, and invention, Popular Mechanics and the History Channel assembled a panel of experts who helped select and rank the 150 most significant gadgets.

Buzz Aldrin
Apollo 11 astronaut

Greg Alwood
Director, Children's Safe Drinking Water at Proctor & Gamble

Paola Antonelli
Senior curator of architecture and design, Museum of Modern Art, New York

Bianca Bosker
Executive Technology Editor, Huffington Post

George Davison
Founder and CEO of Davison International

James Dyson
Inventor, Dyson Dual Cyclone vacuum

Shawn Frayne
Inventor/president, Humdinger Wind Energy

Lauren Goode
Former producer/reporter, The Wall Street Journal Digital

Lonnie Johnson
Inventor, Super Soaker

Brian Lam
Former Editorial Director, Gizmodo

Tim Leatherman
Inventor/founder, Leatherman Tool Group

John Maeda
President, Rhode Island School of Design

David Pogue
Technology columnist, The New York Times

Elspeth Rountree
Digital strategist

Witold Rybczynski
Martin & Margy Meyerson Professor Emeritus of Urbanism and Director of real estate, The Wharton School, University of Pennsylvania

Tim Wu
Professor of law, Columbia University; author, The Master Switch

150 | REMOTE CONTROL KEY FOB

INVENTOR RENAULT ENGINEERS | **1982**

Keyless car entry used to mean a rock through a window. But, with increased frequency since the 1980s, a click of a key fob can get you into a vehicle using a computerized system of rolling codes designed to ensure that your vehicle doesn't respond to anyone else's clicks.

149 BOWFLEX

INVENTOR TESSEMA SHIFFERAW | **1984**

WHO WOULD IMAGINE THAT A REVOLUTION IN ETHIOPIA WOULD HAVE A PROFOUND INFLUENCE ON THE AMERICAN HOME EXERCISE MARKET?

Tessema Dosho Shifferaw was an Ethiopian student studying industrial design at San Francisco City College when his country's government was overthrown and its leader, Emperor Haile Selassie I,

imprisoned in 1974. Shifferaw's father, an army general, was later executed. Stranded in the U.S., Shifferaw continued his studies, driving a cab to pay his way through school. While trying to develop an ergonomic chair for a class project in 1984, he invented the Bowflex exercise system. Though other systems used resistance training, Shifferaw's variation used bendable rods instead of weight plates. After a few iterations of the product—and

Another benefit of Bowflex Power Rods: less soreness than with free weights, because muscles don't have to work as hard to lengthen back to a starting position.

a successful stock offering— Bowflex began pushing hard with TV infomercials in 1996, focusing sales on direct marketing. In addition to the exercise benefits, its compactness and zero-down financing were major selling points.

VIEW-MASTER

INVENTOR **WILHELM GRUBER** | 1939

t's probably difficult for anyone born after 1980 to imagine what entertainment value there could possibly be in slowly viewing seven 3D images through an awkward plastic device. But click we did—with the View-Master, **a stereoscopic device first sold in photo shops that then transitioned into a toy store bestseller.** At the peak of its popularity, the View-Master was hawked on TV by movie star Henry Fonda.

The real master of the View-Master, though, was German immigrant Wilhelm Gruber, an organ maker and piano tuner who was spotted by Harold Graves, the president of postcard company Sawyer's Inc., in 1939 as he set up a 3D photograph using two cameras. The innovation: Pairing the images on a disc which, when looked at through the right device, would pull them together into a coming-at-you picture (see 3D Glasses, page 16). Sawyer's introduced the new device at the 1939 New York World's Fair.

The U.S. military was the bread and butter for View-Master through World War II until the company bought a competitor, Tru-Vue, which

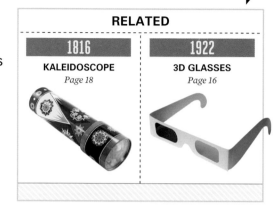

RELATED

1816	1922
KALEIDOSCOPE *Page 18*	**3D GLASSES** *Page 16*

had a lucrative licensing deal with Disney. By the time General Airline & Film bought the company in 1966, it had largely refocused to children's content—images that live on in the memories of fans today.

BUNGEE CORD
INVENTOR VARIOUS | DATE UNKNOWN

ALSO KNOWN AS SHOCK CORDS, bungee (or bungy) cords have become essential elements of camping trips, mountain expeditions, military excursions, and college move-ins.

While the springy little do-it-alls have been in use since before World War I, it's unclear who first surrounded elastic, rubbery material with flexible fabric, put hooks on the ends, and, in the process, figured out a way to pack and secure objects without tying knots. We don't even know for sure who was the first person to strap a cord to his or her ankles and leap from a high place—although we do know that the Oxford Dangerous Sports Club took to the skies in England in 1979, leaping off the Clifton Suspension Bridge in Bristol, England, in the first recorded bungee jump in history.

WHAT A HACKETT
The man responsible for turning this handy packing device into the core technology of an extreme sport is New Zealander A.J. Hackett. He leapt from the Eiffel Tower in 1987, without permission, using a super-stretchy bungee of his own design.

146 RETRACTABLE DOG LEASH
INVENTOR MARY A. DELANEY | 1908

Though now linked to many dog and human injuries, the retractable dog leash originally was intended to keep animals under control while giving them comfort and the freedom to roam. New York City dog owner Mary A. Delaney patented the first retractable leading device, which attached to a dog's collar, in 1908. Later versions, including Billy G. Crutchfield's 1978 invention, attached the long leash to a plastic handle, so dogs could be reeled back in fishing rod-style.

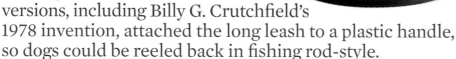

145 POWERMAT
INVENTOR RAN POLIAKINE | 2009

FROM CELLPHONES TO LAPTOPS and dry-cell powered flashlights, we live in a world of rechargeables. That's great and green, but there's a downside: a plethora of chargers and their associated cables. Enter the Powermat, a relative newcomer in the gadget department. It avoids the tangle by going wireless and using inductive energy transfer. The charging base station and device work together as a transformer. The primary induction coil within the charging base station creates an alternating electromagnetic field, while the second induction coil in a handheld device takes that power and converts it into an electrical charge to power its battery.

144 AIR MATTRESS

INVENTOR VARIOUS | 1903

IN ANCIENT EGYPT commoners slept on piles of palm fronds. In Rome mattresses were stuffed with hay and wool. During the Renaissance peashucks were used as stuffing.

But while cast-iron frames and coil springs were great innovations neither was much help when it came to a) camping and b) accommodating the needs of a visiting brother-in-law in a house without a dedicated guest room bed.

Campers and in-laws weren't the only customers drawn to the pump-and-sleep technology. Steamship lines were early adopters, believing that air mattresses were valuable not just for comfort and roll-up convenience, but also for their use as flotation devices should disaster strike. Imagine—if Jack had had an air mattress, *Titanic* might have ended quite differently.

METZELER *Camping*

ZELT · UND SCHWIMMATRATZEN · LUFTBOOT

Pressed garlic has a lighter flavor because it removes the bitter center stem. Outspoken chef /TV personality Anthony Bourdain doesn't approve: "I don't know what that junk is that squeezes out of the end of those things, but it ain't garlic."

YOU COULD JUST CHOP your garlic, peeling it first, of course, and then make big pieces into increasingly smaller pieces by carefully wielding a knife. That's how it's been done since the time of ancient Egyptians.

But tedious chopping isn't nearly as much fun as watching garlic forced by pressure through a grid of tiny holes like aromatic Play-Doh. Inventor Steven O. Sarossy patented the garlic press in 1957, later inspiring the cranking garlic chopper.

It's hard to believe that this handy kitchen gadget didn't really catch on until the last half of the 20th century. And while the debate rages on as to whether this convenience serves or sacrifices the flavor, few can argue that the garlic press is one of our kitchens' most entertaining gadgets.

142 VEGETABLE PEELER
INVENTOR VARIOUS | DATE UNKNOWN

HISTORY HASN'T FOUND THE EUREKA MOMENT or even the inspired inventor responsible for the creation of a device that helps remove the outerwear from cucumbers, carrots, apples, and their healthy brethren.

Whatever its origins, the humble vegetable peeler has been tweaked and tampered with regularly since the 1800s, as culinary technicians attempt to upgrade what already works. Most varieties fall into two camps, the Yorkshire, or Lancaster, peeler (characterized by a blade extending out from a handle) and the Y-peeler, in which the blade crosses the handle, as with a shaving razor. But no matter how many infomercials tout the next generation of peelers (the "Titan," for instance, shaves on both the upstroke *and* the downstroke), the challenge of actually improving the device speaks to the essential rightness of the gadget itself.

141 TIRE-PRESSURE GAUGE
INVENTOR PETER W. PFEIFFER | 1927

PROPERLY INFLATED TIRES, as you no doubt have heard from a nagging parent or well-meaning mechanic, increase the life of your tires and improve the overall mileage of your vehicle. Still, if you have a tire gauge, there's a good chance it's buried in your glove compartment underneath your owner's manual.

The technology of the tire-pressure gauge, filed by Peter W. Pfeiffer in 1927, is simple. By attaching this handy device to a tire and creating a solid seal, the air coming out of the tire moves a piston that's attached to a spring for resistance.

The more pressure, the farther the piston moves, and the higher the reading you see. Variations include gauges with circular readouts and, more often these days, built-in pressure sensors that can be read on the dashboard.

With such advances, is the humble piston gauge on the way out? The well-worn piston design may go the way of its first partner, the Ford Model T. But like the automobile itself, it seems the pressure gauge is a gadget that's here to stay.

(140) | PORTABLE BICYCLE PUMP

INVENTOR VARIOUS | 1897

EARLY BICYCLES used all-rubber tires—which didn't provide the smoothest rides. Enter Scottish veterinary surgeon, John Boyd Dunlop, who introduced the pneumatic tire in 1888 after experimenting with a garden hose.

But a tube needs a pump, or it's just a tube rather than a tire. Similar to a tire pressure gauge but working in reverse, a tire pump forces air into the tire's inner tube. In 1892 Englishman Gordon Rushbrooke got the idea for a pump mechanism while cleaning an exhaust pipe. His brother Frederick established Halfords ironmongery to begin manufacturing and selling the pump.

In 1905, William S. Feel of Michigan patented a combination spring and pneumatic cushion that supported the seat of the bike and pump and could also be converted into a pump. A variety of other pumps have followed—floor pumps, hand pumps, and electric versions with high and low air pressure modes, to name a few.

RELATED

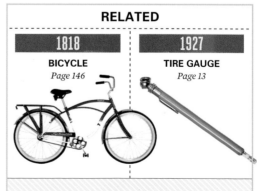

1818	1927
BICYCLE	**TIRE GAUGE**
Page 146	*Page 13*

INVENTOR EDWARD HAUCK | 1884

James Bond wanted his shaken, not stirred. But the idea for the cocktail shaker—simply using a pair of tumblers to mix ingredients in a drink—goes back much further than Ian Fleming's superspy.

Gourds were probably the first such devices. But the modern cocktail shakers evolved through a series of patents from the 1870s to Prohibition. The first ones resembled teapots, since the cocktail hour piggybacked on teatime. Edward Hauck patented the first three-piece shaker with a built-in strainer and an air vent for faster pouring in 1884. **The martini boom in the mid-20th century made a cocktail shaker a must-have for every home bar,** with interest peaking after Prohibition when Hollywood icons mixed drinks on-screen.

Which brings us back to Bond. There's a good reason to have your cocktail shaken instead of stirred, but it's not what you think. The ice in a shaken drink melts more quickly, thus diluting the alcohol. Maybe 007 was trying to pace himself.

The bartending rule of thumb is to shake cocktails with juice, dairy, or egg whites and stir cocktails that only use spirits.

THE MOVIE BUSINESS was chugging along just fine in the first half of the 20th century. But then along came television, bringing visual entertainment into the home. Something was needed, or so thought Hollywood hucksters, to get people out of their living rooms and back into theaters.

The technology for 3D wasn't brand-new. Stereoscope photographs go back to the mid-1800s, and in 1915 Edwin S. Porter, director of *The Great Train Robbery*, was experimenting with ways to bring his films off the screen and into the laps of audiences. But the early 1950s was when the novelty boomed into a phenomenon, with 1953's *Bwana Devil* leading the way. Serious directors took a shot, too, particularly Alfred Hitchcock, who shot *Dial M for Murder* in 3D, though the film played in most theaters in 2D, thanks to waning interest in the gimmick by the time the movie was released in 1954.

With the exception of a few outliers, 3D largely disappeared through the 1960s and '70s, emerging occasionally in the '80s in efforts to wring a bit more box office juice out of a lame cash-in sequel

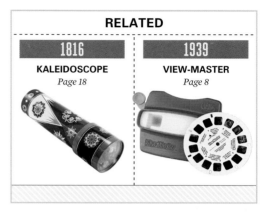

RELATED

1816
KALEIDOSCOPE
Page 18

1939
VIEW-MASTER
Page 8

(*Jaws 3-D*, *Amityville 3-D*, et al.) and to boost attendance at documentaries at IMAX theaters. Then along came director James Cameron, who gave the genre new life by developing technology for his documentary *Ghosts of the Abyss* that he later put to use (or overuse, depending on your critical thoughts) in *Avatar*.

The floodgates opened, not only with new films making use of 3D technology, but with in-development films and cinema classics being retrofitted for in-your-face action. Three-demensional home television isn't far behind—leaving moviemakers once again scrambling for the next hot new gimmick to keep viewers in theater seats. Stay tuned . . .

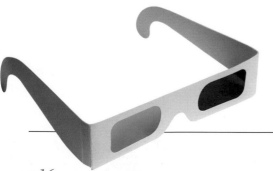

HOW IT WORKS
The red and blue filters on old-school 3D glasses work by feeding two different images—one red and one blue—into each eye. Your brain does the rest.

AT FIRST GLANCE, a kaleidoscope doesn't serve much of a purpose—you look in it, you marvel at the colorful display, you turn the opposite end and watch the colors and shapes slowly change . . . and then you break it open to try to figure out how it works.

If you did the last part carefully, you learned that, like many a magic trick, **the kaleidoscope's wonder is pulled off with mirrors.** Presented at the proper

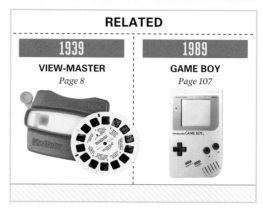

RELATED

1939
VIEW-MASTER
Page 8

1989
GAME BOY
Page 107

angle, combined with small objects (especially shiny ones), and encased in a spyglass-like tube, the kaleidoscope creates the illusion that you are seeing a cross section of a glass-filled tube when you are, in fact, only seeing one wedge-shaped section of it reflected.

Though mirrors had been used decoratively and even by magicians to create illusions, it wasn't until 1816 that Dr. David Brewster of Scotland positioned them in a tube as an aid to artists, whom he hoped would be inspired by the patterns therein, thus inventing the kaleidoscope and its endless variations.

ICE CREAM SCOOP

INVENTOR ALFRED L. CRALLE | 1896

Many people complained that ice cream stuck to spoons and usually required two hands to get from container to bowl. But Alfred L. Cralle, a Pittsburgh drugstore porter, did something about it. In 1897 he patented a gadget that used a squeezable handle to push the cold confection out of its attached cup. The "Ice Cream Mold and Disher" is still widely used today, but we call it the ice cream scoop.

RADAR DETECTOR

INVENTOR RADATRON, INC. | 1962

Before there were radar detectors, there was radar. An acronym for RAdio Detecting And Ranging, radar transmits radio waves, which bounce back to the sending point. How long that return takes can help determine the location of an object or how quickly that object is moving. Since the 1950s, police have been using the military-created technology to nab lead-foot drivers for exceeding the speed limit.

But for every action, science tells us, there's an equal and opposite reaction. In this case the opposite reaction was the development of the radar detector. A number of versions began appearing in the early '60s, including the Sentry, which *Time* magazine stated was "almost useless" at high speeds.

The defining brand came in the late '60s when Dale Smith of Dayton, Ohio, created the Fuzzbuster, billed as a "parametric X-band radar receiver." Translation: It alerted drivers to the operation of police radar. Even as other companies entered the market—and legislation against them was pushed through lawmaking bodies— the brand name Fuzzbuster became the generic term for all radar-dodging devices.

FUZZ BUSTED

Under the Federal Communications Act of 1934, most private U.S. vehicles are allowed to have radar detectors, while most commercial vehicles are not. The penalty for having a verboten radar? Confiscation and a ticket, which you were trying to avoid in the first place.

YOU COULD MAKE A CASE FOR THE BATHROOM SCALE BEING ONE OF THE MOST COMMON HEALTH CARE PRODUCTS IN THE HOME. BUT THE DEVICE IS LARGELY A 20TH-CENTURY ADDITION.

Prior to that, the scale was a coin-operated novelty brought to the U.S. from Germany in the 1880s. Even into the 1920s and '30s, it was common practice to drop a penny in a public machine crafted by the Peerless Scale Company to find out your weight, sometimes even receiving a printed fortune along with it. But as the home bathroom scale emerged in the 1940s, the public scale was (thankfully) phased out.

The basic principle behind the body-weight scale is a series of levers that distributes your weight evenly and stretches a main spring. The more it stretches, the heavier you are—and the more it turns a dial to pinpoint your poundage. Later, electronic versions used a strain gauge and a load cell sensor to translate the force being exerted on the scale into an electronic signal. The result? Not always what you want it to be.

RELATED

Circa 1500

PEDOMETER
Page 29

▶ To be fair, an anonymous Inca in about the year 300 should get credit for creating the popcorn maker. Colonists in what would later be the United States were also popping kernels over a fire. But the acclaim for modern popcorn makers goes to Charles Cretors, a peanut and candy purveyor who, with the help of a mechanical clown, The Toasty Roasty Man, popped popcorn for hungry crowds at the World's Columbian Exposition of 1893 in Chicago. Soon he had crafted a horse-drawn popcorn wagon and was experimenting with electric motors to spark the process. The result: The Earnmore popper—which proved popular with street vendors hawking their buttery wares. The Great Depression helped the business, as snackers looked for a low-cost alternative to expensive sugar-based candies and found it in popcorn.

Cretors' creation and other popcorn makers work on the same principle: Steam or oil is used to heat the kernel, whose interior starch builds pressure until it erupts through the harder hull. On their way out, the starch and proteins form the familiar white abstract shape of the popped corn.

None of which matters, of course, as you're shoveling handfuls at a time into your mouth while watching a movie.

WHAT'S POPPIN'?

Charles Cretors harnessed the latest technology to sell his popcorn. He started by upgrading an old peanut roaster in 1885, emphasizing its high-tech steam engine. Then in 1900 Cretors invented the first popcorn machine with an electric motor, earning one of the oldest active Underwriter Laboratory numbers (EA4175) for electrically operated machinery.

132 LASER POINTER

INVENTOR VARIOUS | 1980s

SUITABLE FOR BOTH INDICATING KEY POINTS IN A COLLEGE LECTURE AND ENTERTAINING YOUR CATS,

the laser pointer is a handy, scaled down version of the laser (an acronym for Light Amplification by Stimulating Emission of Radiation, which pretty much tells you how it works).

The laser itself was developed by, well, that's debatable. It could be physicist Theodore Maiman or Nobel Prize winners Charles Townes or Arthur Schawlow. Like computers, lasers were expensive propositions when first crafted—and primarily used by the military and James Bond villains. But by the 1980s, the cost had been lowered enough to allow for such inexpensive, if originally bulky, items as personal laser pointers, which focused controlled red beams produced by a diode. Green, blue, and yellow lasers followed, each of which uses a different diode.

DR. NO WE'RE NOT, but, whatever the color, the laser pointer still reigns supreme as one of our favorite silly gadgets of the century.

131 PAINT ROLLER

INVENTOR RICHARD C. ADAMS | 1942

IT WASN'T UNTIL THE 19TH CENTURY that the painting of homes was common practice. But between the practice becoming acceptable and the invention of the paint roller in 1940, slapping a new coat of paint onto a wall or the side of a house meant painstakingly brushing stroke by stroke.

With a roller, though, painting is, well, maybe not easy, but at least easier. And it's harder to identify the specific spot where you got tired.

Whom you credit with the creation of the paint roller could depend, in part, on your national loyalties. In 1940, Canadian Norman Breakey was the first to create a basic version of the device. However, American Richard Croxton Adams—a descendant of John Adams—earned the first patent while working a day job with Sherwin-Williams.

130 LABEL MAKER

INVENTOR R. STANTON AVERY | 1935

R. STANTON AVERY created the first self-adhesive label-making machine in 1935 using, among other things, a motor from a washing machine. At the time his company was called Kum Kleen Productions, going through a series of name changes before becoming the multi-billion-dollar grossing Avery Dennison Corporation.

The first personal label maker, called an embosser, was sold by DYMO in 1958. Early embossers required the turning of a dial to get the proper character in place and then squeezing a handle grip to imprint the hard plastic adhesive tape strip. Later electronic versions featured LCD screens and functioned more like printers. Some even talk.

While the advent of inkjet computer printing has affected the label maker, it hasn't destroyed it. And the obsessively organized among us have R. Stanton Avery to thank for the labeled boxes of old cassettes and tax records deep in our closets.

No denying, pepper spray is nasty stuff. The extract of dried chili pepper fruit, it's chemically known as Oleoresin Capsicum (OC) and is designed to make the eyes of an attacker so teary and painful that attacking someone becomes a very low priority.

As an inflammatory, its immediate effects include coughing, difficulty breathing, and the inability to keep the eyes open, all lasting an average of 45 minutes, depending on how much has been sprayed and how directly.

Inventor Kamran Loghman-Adham not only helped develop the spray, which had been used in smaller doses, as a dog repellent, but also developed usage procedures for police. To his dismay, his invention hasn't always been used as intended.

Pepper spray inventor Kamran Loghman-Adham told a reporter, "It is becoming more and more fashionable . . . to use the chemical on people who have an opinion . . . It's not a thing that solves any problems nor is it something that quiets people down."

Pictured, West Virginia University students directly hit with pepper spray following the upset win of West Virginia over Virginia Tech in 2003.

128 TASER

INVENTOR JOHN H. "JACK" COVER | 1972

A MAN WALKED INTO AN ELECTRIC FENCE AND SURVIVED.

Who that unfortunate fellow was, we don't know. But we do know that reading about the incident inspired NASA researcher Jack Cover. If electric current could be debilitating without being deadly, a device that controlled such force might be a boon to law enforcement.

And so the TASER—a semiacronym for Tom Swift Electric Rifle, in tribute to the kid fiction hero—was born. With a TASER, a tethered dart is shot at the target, sending a high-voltage/low-amp dose of electricity that interferes with the brain's communication to the body's muscles. The folks at TASER International compare it with the static on a telephone line: "Once the static stops, communication continues and there's no damage to the phone."

In the meantime, though, the phone sure gets one hell of a sting.

STUN ON THE RUN

Modern police TASERs now include the controversial Drive Stun, also known as Dry Stun, capability. The device is held against the body to cause pain without affecting the central nervous system and incapacitating the target.

SCOOTER

INVENTOR VARIOUS | DATE UNKNOWN

▶ Essentially a skateboard with a steering column, the scooter (or kick scooter, to differentiate it from motorized scooters) is a low-tech method of transport that has gone in and out of popularity since kids were crafting them out of discarded wooden orange crates during the Great Depression.

The Razor Scooter, a more contemporary, foldable, aluminum variant, was not the first collapsible scooter in existence. Credit goes to a number of sources, including Swiss banker Wim Ouboter, who set out to design a scooter that could fold into his backpack to avoid embarrassment.

But the one that turned scooters into a worldwide phenomenon was the work of Taiwan's J.D. Corporation and engineer Gino Tsaim. Tsaim claims he wasn't thinking about the mass market when he invented the Razor Scooter. He was merely answering his boss's call for a way to scoot around the bicycle factory.

126 PAGER

INVENTOR ALFRED J. GROSS | 1950

PEOPLE WERE BEING PAGED long before there were pagers. The verb simply referred to efforts to publicly track someone down, whether a doctor in a hospital or a stray passenger at an airport.

But the actual gadget, the personal pager, was an essential belt accessory if you were a doctor, CEO, repairman, or drug dealer—before cellphone technology pretty much put it on the heap that includes the 8-track tape player and the CB radio.

The creation of Al Gross, the man also largely responsible for the walkie-talkie, the pager was designed for doctors and first put to use at New York's Jewish Hospital in 1950. It didn't hit the commercial market until the 1970s, and early models didn't transmit messages. Using a dedicated radio frequency, it only notified the receiver that there was a message.

Sound limiting? Then you weren't paying attention in 1998, when pagers were at the peak of their popularity. Motorola was the biggest player, controlling 80 percent of the market.

125 PEDOMETER

INVENTOR LEONARDO DA VINCI | CIRCA 1500

Counting steps isn't a modern invention. Thomas Jefferson had an interest in the subject, as did Leonardo da Vinci, whose *Codex Atlanticus* sketchbook included a plan for a pendulum-based device that mimics a walker's gait.

RELATED

1931

BATHROOM SCALE
Page 21

29 •

Put a couple of roller skates on a board and what do you get?

Bruised knees, for one. You also get a device that allows for sidewalk surfing, if you can stay on it. Surfers used the first skateboards to ride the streets the way they rode the waves.

The fad lasted until the mid-'60s, and then lost momentum.

The business got a boost in 1973 with the introduction of urethane wheels that cut down the noise and bumps that came with the original clay wheels. The California drought of 1976 and 1977, which forced the draining of backyard pools, opened up new vistas for daredevil skateboarders. And then a legend was born, as Alan "Ollie" Gelfand figured out that you could slam a foot down on the skateboard tail while jumping and launch both you and the board airborne.

By the time ESPN broadcast its first Extreme Games in 1995, the former toy had been widely accepted as a bona fide sporting tool by mainstream audiences. Not bad for a bunch of surfers and long-haired punks.

▶ **Skateboarding kicked off with larger-wheeled longboards and evolved to include shorter, lighter street boards for performing—or attempting—tricks.**

123 | CURLING IRON
INVENTOR HIRAM MAXIM | 1866

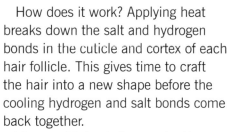

INVENTORS GONNA INVENT. Which is why it's not so unusual that Hiram Maxim, the man who designed a machine gun and gun silencer, also held a patent for a curling iron . . . and then went on to create a headlamp for locomotives.

Of course, Maxim wasn't the first to alter the structure of hair with heat. People have been doing that since before the discovery of electricity. Ancient Persians, Babylonians, and Assyrians were among those who heated and shaped their coifs.

How does it work? Applying heat breaks down the salt and hydrogen bonds in the cuticle and cortex of each hair follicle. This gives time to craft the hair into a new shape before the cooling hydrogen and salt bonds come back together.

The result? A whole new do, if your hair can hold the curl.

122 NAIL CLIPPER
INVENTOR VALENTINE FOGERTY | 1875

BEFORE THE INVENTION OF THE NAIL CLIPPER, you had three basic choices when dealing with the inevitable growth of your fingernails. You could let them grow. You could use scissors to trim them. Or you could use your teeth . . . but nobody wanted to see that.

It's debatable when exactly that changed, although a series of patents for fingernail trimmers were granted between 1875 and the turn of the century, around the time the H.T. Cook Machine Company produced the familiar Gem brand.

While there are many variations, the most common structure is a four-piece metal gadget in which the rotatable top piece can be pushed down to force the two others to clench together. The addition of files and pointed edges makes the tool useful for cleaning and buffing as well as clipping. What it doesn't help you with is tracking down that stray nail shard that flicked out of sight.

121 ROLLERBLADES

INVENTOR SCOTT AND BRENNAN OLSON | 1982

THE HISTORY OF ROLLERBLADES weaves in and around the history of roller skates. The first roller skates, beginning in the 18th century, were in-line versions, with the four-wheels-per-foot version being developed a century later.

Rollerblades were the creation of Scott and Brennan Olson, a pair of hockey-playing brothers from Minnesota who found a pair of older in-line skates and were inspired. They switched out the blades in hockey skates with wheels, crafted a rubber brake for the heel, and concocted a way to practice in the ice-free off-season. Their Rollerblade brand was bought out in 1983 before in-line skating became an iconic sight, particularly in the beach towns of Southern California.

Since then, in-line technology has evolved to meet the needs of four distinct markets: hockey players, recreational/fitness users, speed skaters, and ramp-jumping tricksters, with different wheels, ball bearings, and even lubricants preferred by each.

RELATED

1863

ROLLERSKATES
Page 34

120 | DESK FAN

INVENTOR SCHUYLER SKAATS WHEELER | 1882

▶ Schuyler Skaats Wheeler, a 22-year-old New York native, took Edison and Tesla's breakthroughs in electricity and put them to the service of cooling.

Formerly on Edison's engineering staff, Wheeler developed the first desktop electric fan, taking it to the marketplace in 1882.

That early model wouldn't have passed even basic safety codes today: There was nothing protecting unsuspecting fingers from the rotating pair of blades. But like all fans, it doesn't really cool the air; it just creates windchill. Moving air helps evaporate sweat, leaving you feeling cooler.

119 ROLLER-SKATES

INVENTOR JAMES LEONARD PLIMPTON | 1863

SHOES PLUS WHEELS EQUALS FUN, RIGHT? The first to figure out that equation was John Joseph Merlin, a Belgian-born Londoner who put the two elements together in the 1760s, and is said to have worn his creations to a costume party at Carlisle House where he injured himself and others by crashing into a mirror.

Then in 1819 a Frenchman by the curious name of Monsieur Petitbled patented a wooden-soled inline skate with leather straps instead of a full boot. The only problem: The skates only went forward.

James Leonard Plimpton, considered the father of modern rollerskating, had greater success in 1863. His primary innovation: Putting four wheels on two axles (previous versions were in-line) and attaching the wheels to a rubber-cushioned mobile axle, instead of the bottom of the skate, which have skaters the ability to steer by tilting their bodies.

Plimpton wasn't just a tinkerer, he was a skating visionary. He transformed part of his furniture business pace into a makeshift roller rink and rented skates, and also founded the New York Roller Skating Association to push the new recreational activity.

RELATED

1982

ROLLERBLADES
Page 32

Just three years after inventing rollerskates, James Plimpton opened America's first public roller rink in Newport, R.I.

It helped set off a phenomenon—and sparked a new sport, roller derby (pictured).

118 | DRY-ERASE BOARD/MARKER
INVENTOR JAMES E. ARBERRY | 1939

Thanks to a 1991 sale to Kmart—followed by Walmart, Target, and more—Michael Boone became the dry-erase board guy. At his peak he sold 30,000 a day.

SOMETIMES IT ISN'T THE PIONEER WHO GETS CREDIT. Instead, it's the guy who first slaps his name on a product. And so, while versions of dry-erase boards (aka white boards, as opposed to blackboards) were being developed for years, Michael Boone's Boone Boards were the first targeted to the consumer market.

Pittsburgh inventor James E. Arberry patented the whiteboard in 1940, citing its light weight, durability, easy-to-clean surface, and pleasing color as advantages over traditional slate. The first whiteboards marketed in the 1960s used wet-wipe markers, so they didn't stay white for long. Pilot Pen patented the first dry-erase marker, invented by Jerry Woolf, in 1975.

117 | ROLLING SUITCASE
INVENTOR BERNARD SADOW | 1970

THIS SEEMINGLY OBVIOUS BREAKTHROUGH in traveler mobility didn't happen until 1970, when Bernard D. Sadow and his family had to carry heavy suitcases through the airport after a vacation in Aruba. While waiting at customs, he observed an airport worker moving a machine on a wheeled skid steer loader. "I said to my wife, 'You know, that's what we need for luggage,'" Sadow later recalled. **He attached casters to his luggage when he got home and patented the idea two years later.** Even then it faced resistance in the marketplace.

Why did it take so long? One theory is that people didn't have to carry luggage around that much—or at least, the luggage owners didn't. Before the post WWII aviation boom, trains and boats were the primary means of long-distance transportation—which meant that porters were primarily responsible for the heavy lifting.

Luggage innovation doesn't travel very quickly. After the 1970 invention of the rolling suitcase, it took nearly 17 years to introduce the telescoping handle.

Fig. 1. — Trépidation du front (Liedbeck).

A LESS PERSONAL MASSAGER
The Hammer made Dr. Granville's work much more efficient. He'd previously treated hysteria patients by hand.

LET'S NOT GET INTO Cleopatra's apocryphal gourd full of bees or the hand-cranked and steam-powered experiments from the Victorian era. Instead, we'll join the makers of the film *Hysteria* in giving credit to Dr. Joseph Mortimer Granville, who was the first to patent an electromechanical vibrator in the 1880s. Nicknamed The Hammer, it looked like a drill with a ball on the tip.

At the time, the device wasn't used as a sexual stimulant. It was intended to treat "hysterics"—women who supposedly suffered from the "stagnation of the female juices." Others saw it as a treatment for arthritis, constipation, hair loss, and more. Pre-1930s advertisements touted them as muscle relaxants.

Now of course, we know better. And a seemingly infinite number of variations of the device are on the market.

SCUBA

INVENTORS JACQUES-YVES COUSTEAU, EMILE GAGNAN | 1946

UNDERWATER ADVENTURES were once limited to the amount of time you could travel round trip while holding your breath. SCUBA—an acronym for Self-Contained Underwater Breathing Apparatus, coined by U.S. Army doctor Major Christian Lambertsen—changed that.

Which is not to say that Lambertsen invented SCUBA gear. In the early 1940s he developed the closed-circuit rebreather for the U.S. Navy, which absorbs carbon dioxide from exhaled breath to allow rebreathing of unused oxygen.

In 1946, famed oceanographer Jacques-Yves Cousteau (left) and creative partner Emile Gagnan earned a spot in the National Inventors Hall of Fame by developing the **first open-circuit system in which gas flowed from the cylinder to the diver and out into the water. It compressed air while also making adjustments for the surrounding pressure,** allowing deep sea diving. The Aqua-Lung went on the market in 1952.

114 | MODERN COMBINATION LOCK
INVENTOR JOSEPH LOCH | 1866

THE BASIC MODERN COMBINATION LOCK works by attaching a combination dial to a spindle that controls a series of wheels.

Al-Jazari, author of *Book of Knowledge of Ingenious Mechanical Devices*, provided plans for a combination lock—along with a mechanical clock powered by water and weights—in 1205 A.D. But German inventor Joseph Loch is usually cited as the man who crafted the first modern iteration for Tiffany & Co., which had plenty worth locking up.

When the notches in each wheel line up with the others, the bolt that secures the lock is released.

113 MIXMASTER
INVENTOR IVAR JEPSON | 1928

Brought life to meringues across America.

WHEN IT COMES TO MIXING THINGS UP in the kitchen, you've probably heard the names Hamilton and Beach. Perhaps if Sunbeam had been renamed Jepson Industries, the man responsible for the Mixmaster mixer would get the recognition that he, too, deserves.

Ivar Jepson was the head designer for Chicago Flexible Shaft Company, a manufacturer formerly known for its driveshafts and mechanical sheep shears, when he developed the iconic two-beater Mixmaster for the company's Sunbeam division in 1928.

The Swedish immigrant was also responsible for innovations in, among many other things, shavers, waffle irons, and boat hoists. But it was his affordable Mixmaster mixer—with its detachable beaters and pivot arm—that helped make Sunbeam a household name.

ELECTRIC IRON

INVENTOR HENRY SEELEY | 1881

ELECTRICALLY, AN IRON IS SIMPLE;

its circuit consists of cord, thermostat, and heating element. The string tension on the thermostat elements is varied to give different temperature settings, and the iron, once hot, will cycle on and off within a few degrees of the desired temperature. The goal: For heat to loosen long-chain polymer molecules in fibers, while the weight of the iron straightens them out.

But the chore of ironing pre-dates any understanding of "long-chain polymers." Indeed, the popular curve-and-point shape of most irons dates back to before the electrical days. (The point proved useful in dodging buttons.) These early irons had to be heated on wood stoves before use.

In 1882 the first electric iron was patented by Henry W. Seeley, using a carbon arc for heat—not high on the safety scale. A decade later Crompton and Co. and the General Electric Company improved it with a version of the technology used today. The iron soon became the first mass-market electric appliance, finding its way into more than 90% of American homes today.

111 TAPE DISPENSER

INVENTOR JOHN A. BORDEN | 1932

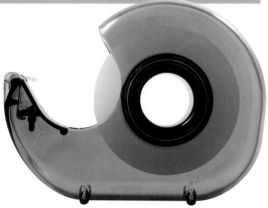

SYMBIOTICALLY ADHERED TO EACH OTHER in consumer invention history, Richard Drew and John A. Borden were both 3M engineers when they respectively changed the face of office desktops and craft projects.

3M (which stood for Minnesota Mining and Manufacturing) was primarily a sandpaper concern when Drew introduced what later became known as Scotch tape back in the 1930s. The combination of coated cellophane and adhesive appealed to the public, but Borden really made it stick by designing a practical dispenser with a built-in blade to cut the tape. In 1932 the first dispenser on the market weighed nearly seven pounds.

110 HOME THERMOMETER

INVENTOR GABRIEL FAHRENHEIT | 1714

Let's take this one a step (or two) at a time. First there was the thermoscope (commonly known as the Galileo thermometer, after its inventor) in 1593, a device that could indicate a change in temperature but not quantify it. A scale of measurement wasn't included until 1612, when Santorio Santorii developed his air thermoscope. Mercury didn't become part of the mix until 1714, when Daniel Gabriel Fahrenheit put it in a thermometer. In 1724 he introduced his now widely adopted temperature scale based on the freezing point of water.

Electric thermometers bypass the liquid and use a sensor—a thermoresistor—that changes its resistance depending on the temperature.

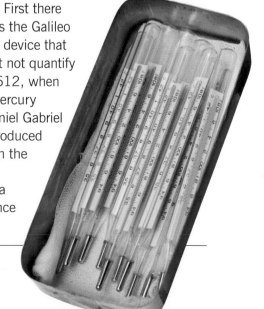

BABY CAR SEAT

INVENTOR LEONARD RIVKIN | 1962

A h, the unsafe old days, when a family could load into a station wagon without regard for the safety of the children climbing from seat to seat without a strap in sight. Babies were particularly vulnerable—at best they were strapped into a backpack-like sack that hung on the back of a seat.

Leonard Rivkin, a Colorado furniture store founder, is credited with the creation of car seats in 1963. Ford began offering them as options in 1965, recommending that parents let kids play with them at home to get used to the confinement before heading out on the road. (If this seems late in the game, consider that seatbelts weren't standard equipment in U.S. cars until 1964.) Pediatrician Seymour Charles became the Ralph Nader of car seat safety when he created Physicians for Automotive Safety in 1965 and began pushing for government regulation of car seat and seatbelt laws. Government standards were set in 1971, and by the following year there were enough models for *Consumer Reports* to start rating them. Legislation started passing in 1977— with exceptions being made for "babes in arms." By 1985 every state had some sort of car seat law.

Having car seats and using them properly, however, proved two different things: A 2006 study showed that 72 percent of car seats were being used improperly.

THE OLD ORDER

A few car seats were manufactured as far back as the 1930s, but they didn't protect children from car accidents. Instead, they were designed to boost little kids up so parents could easily see them in the rearview mirror.

▶ Powerful as they are, smartphones are not truly capable of communication on their own accord. But when the apocalypse comes, we may still be able to use walkie-talkies.

The seemingly archaic devices—which were much cooler before cellphones came around—are really two-way radio transceivers. Unlike phones, you have to take turns with walkie-talkies, pressing a button when you want to talk.

Who's responsible for the gadget we know today? Probably Motorola engineers, with an assist from Canadian inventor Donald Hings and American inventor Al Gross, also responsible for the pager (see page 29). Hings invented a portable 12-pound, 130-mile range radio signaling system that he called a packset in 1937. When Canada declared war on Germany in 1939, he was sent to Ottawa to further develop his packset. Hings said the name walkie-talkie came about in 1941 when a reporter asked what the device did. "Well, you can talk with it, while you walk with it," Hings answered.

Gross worked on the walkie-talkie from 1934 to 1941, which Motorola improved into the handheld "Handie-Talkie." (Previous versions were carried on soldiers' backs.) The device used a vacuum tube and high voltage dry cell batteries. Initially the walkie-talkie was a military device, but soon other uses became clear—pre-cellular search and rescue, marine communication, police work, and as simplified toys.

GIZMO EXPOS

SINCE THE FIRST GLOBAL EXPOSITION AT LONDON'S CRYSTAL PALACE IN 1851, WORLD'S FAIRS HAVE SHOWCASED FUTURISTIC TECHNOLOGY, INCLUDING GADGETS ON OUR LIST. MARVEL AT THESE AWESOME DEVICES!

1876

CENTENNIAL EXPOSITION, PHILADELPHIA

Remington Typographic Machine, forerunner of the qwerty keyboard typewriter (No. 14)
Bell's telephone (No. 7)

1878

EXPOSITION UNIVERSELLE, PARIS

Thomas Edison's phonograph
(No. 8)

1893

WORLD'S COLUMBIAN EXPOSITION, CHICAGO

Separable fasterner, a predecessor of the zipper (No. 65)

1904
LOUISIANA PURCHASE EXPOSITION, ST. LOUIS

Bissell Parlour
Queen Carpet Cleaner,
a manually powered
predecessor of
the vacuum cleaner
(No. 18)

1933
CENTURY OF PROGRESS INTERNATIONAL EXPOSITION, CHICAGO

Singer debuts the
Featherweight, the
first portable sewing
machine
(No. 24)

1939
NEW YORK WORLD'S FAIR, NEW YORK CITY

RCA gives the first
public demonstration
of television (No. 3)
Carrier showcases the
portable air conditioner
(No. 6)

1982
KNOXVILLE INTERNATIONAL ENERGY EXPOSITION, KNOXVILLE, TENNESSEE

Elo TouchSystems demonstrates touchscreen technology, later incorporated into the smartphone (No. 1)

Even today, whether it's by loss, theft, or fire, having your laptop—along with your irreplaceable files—abruptly vanish is something that few of us would like to contemplate. The obvious solution is to protect your data, then stash it somewhere safe—like on a portable hard drive.

For all its computational wonder, the first gigabyte hard drive wasn't what you'd call portable—the 1980 IBJ 3380 clocked in at 550 pounds.

But hard drives are one of the fastest evolving technologies, with capacity doubling every year or two.

The granddaddy of the device is Rey Johnson, who was put in charge of a West Coast IBM unit in 1952, a time when the common belief was that 17 or 18 computers would be all the market could bear.

Developing a magnetic storage machine was among his crew's first challenges (others included creating a nonimpact printer and a test-scoring machine).

Enter the IBM 2311 in 1964, which could store 7.25 megabytes on a removable disk pack. Like its predecessors, it was far bigger than what we would consider a disk drive today. But unlike the others before it, the 2311 offered an electrical connection, making it possible to plug in to different computers.

EVOLUTION OF DIGITAL STORAGE

1971	1988	1999
FLOPPY DISK	**CD-ROM**	**FLASH DRIVE**
Page 112	*Page 94*	*Page 64*

106 TUPPERWARE

INVENTOR EARL S. TUPPER | 1946

THE FIRST SELLING POINT of Earl Tupper's plastic creations wasn't their "burp"-able lids. It was their tightness and near-unbreakability. His airtight seals—inspired by paint cans—became available in 1946 and were perfect for the refrigerators that had become a staple in nearly every home.

The party—the Tupperware party, that is—started two years later when the product wasn't a big store seller. It needed hands-on demonstration. The timing was great, with the growth of the suburbs and an exponential rise in backyard barbecues. The plastic containers started making serious cash, and soon it was obvious that retail outlets weren't necessary.

By 1992, Tupperware was distributed almost entirely by a home sales team, nearly half of whom held other full-time jobs. Say, are you free next weekend?

RELATED

1935

LUNCHBOX
Page 90

Taken from the Latin word for shade, umbrellas were originally designed for protection from the sun, not from the rain.

Usually seen in ancient artwork making royalty more comfortable, umbrellas were first waterproofed by the Chinese, who waxed their paper versions, thus changing the way we deal with inclement weather.

The difference between an umbrella and a parasol? Not much, actually, though umbrellas were traditionally self-carried and parasols carried by others.

Many inventors contributed to the evolution of the umbrella, but the style we're most familiar with today owes itself to industrialist Samuel Fox, who crafted the steel ribbed frame in place of previous versions, which used wood or whalebone. (However, the nod goes to Frenchman John Gedge's 1852 self-opening prototype.) Eventually, the umbrellas made by his company, Fox Umbrellas Ltd., kept President John F. Kennedy, TV's *The Avengers*, and maybe even you dry.

WHEN IT RAINS, IT POURS

Samuel Fox didn't start out in the umbrella business. In 1842 he converted an old corn mill into the steel works Samuel Fox and Company with the goal of producing wire for textile pins. Fox invented the steel ribbed umbrella frame out of leftover farthingale stays used for women's corsets. The product was such a success that he established Fox Umbrellas Ltd. in 1858.

104 iPAD

INVENTOR APPLE, INC. | 2010

BILL GATES CALLED IT "A NICE READER."

He wasn't the only one skeptical of the iPad when it was introduced to the market by Steve Jobs in 2010. There were already other tablet PCs out there—but most of them used a stylus, which is something Jobs felt was unnecessary. Senior Vice President of Industrial Design, Jonathan Ive, also suggested that Apple ditch the keyboard.

It may not have seemed particularly innovative to some—coming across as a kind of hybrid laptop/iPod/e-reader that couldn't handle Flash and wasn't great for multitaskers. It didn't satisfy an unmet need the way the iPhone did. But **the iPad was a big hit, selling 3 million units in 80 days.**

For all the jokes when the iPad was first released, it actually wasn't the first electronic gadget with that name. In 1999, Texas-based Netpliance released an IPAD (all caps) stripped-down computer that featured a 10-inch flat screen and keyboard.

EVOLUTION OF THE PERSONAL COMPUTER

1968	1975	1982
COMPUTER MOUSE *Page 122*	**PERSONAL COMPUTER** *Page 156*	**LAPTOP** *Page 134*

We don't know who invented the compass. But we know that author Zhu Yu was the first to write about it as a navigation tool in 1117. Prior to that, lodestones—minerals with iron oxide that naturally aligned north/south—were used for fortune-telling.

The modern version of the compass can be credited to English inventor William Sturgeon, who also invented the electromagnet. Both versions were simple but kind of miraculous. Aligned with our planet's magnetic field, a compass always points magnetic north.

Knowing how a compass works is still a requirement for scoring a Boy Scouts' merit badge. But the answer isn't really easy. Sure, a compass is basically a magnet balanced on a tiny pivot point, but scientists still aren't unified in explaining why the earth's magnetic field exists and has the influence that it does.

◄

Of course, magnetic north, located on Prince of Wales Island in northern Canada, isn't the same as true north. It's more than 1,500 miles away from the actual North Pole.

EARLY MICROSCOPES were single-power gadgets—basically fixed versions of the magnifying glass. But in 1590 when Zaccharias Janssen (working with his father, scholars believe) lined a few of these up together and took a look, he crafted a device that would change the world.

Galileo later took up the cause and did his own work. So did Anthony Leeuwenhoek, who was the first to identify bacteria and earned the nickname The Father of Microscopy.

Medicine, ecology, genetics, forensics . . . Just about every branch of science was transformed by the microscope.

101 DUCT TAPE

INVENTOR JOHNSON & JOHNSON | 1942

"HOUSTON, WE HAVE A SOLUTION."

In 1970 the three astronauts aboard Apollo 13 used duct tape and other items to fit the command module's square CO_2 scrubbers into the round receptacles of the lunar module they were using as a lifeboat. It saved their lives. Two years later, it came to the rescue again to mend the moonbuggy's fender.

You might say it's a material, not a gadget. But whatever you call it, when an invention allows NASA astronauts to make repairs in space, the MythBusters to build a boat, the Brookhaven National Laboratory to fix its particle accelerator, and enthusiasts to make prom dresses and wallets, it's time to pay attention.

It's very name, though, is a point of some confusion. Yes, Duck is a brand, but according to best-selling authors Jim and Tim, aka the Duck/Duct Tape Guys, the fowl name actually predates the term duct tape. The item itself was originally nicknamed Duck after the U.S. military recruited Johnson & Johnson to create a waterproof tape in 1942. Only after the war, when it was found to be useful on duct work, was the term duct tape used.

Is it miraculous? Maybe that's overstating it. But it should come as no surprise that in Finland and Sweden, the ubiquitous gray adhesive is known as Jesus tape.

100 FIBERGLASS FISHING ROD

INVENTOR VARIOUS | 1940

Fishing goes back as far as, well, probably as far back as hunger itself.

The first recorded rods, used in ancient China, Greece, and Egypt, were wooden. The fiberglass rod didn't come into existence until the middle of the 20th century, when hostilities in Asia curtailed bamboo imports. Rodmakers—including Shakespeare, Phillipson, and Montague—needed a new material to keep anglers equipped with low-cost, quality tackle, and fiberglass fit the bill.

Today there's ongoing debate in the fishing world about which conditions and potential catches are best suited for fiberglass rods. But that doesn't change how revolutionary they were: Lighter than its predecessors, the fiberglass fishing rod reacted more dramatically to the pull of the prey. It proved more durable, too.

RELATED

1820

SPINCAST FISHING REEL
Page 72

99 | STAPLER

INVENTOR GEORGE McGILL | 1876

AT THE 1876 CENTENNIAL EXPOSITION in Philadelphia such wonders as the telephone, the arm of the then-in-progress Statue of Liberty, and a universal grinding machine wowed the public. Also on display: a creative device for inserting a fastener into paper. Its inventor, George McGill, would go on to craft one of the first paper clips, but in Philly it was an early, patented version of the stapler that brought him notice.

Others contributed over the years, most notably Henry R. Heyl, whose 1877 creation combined insertion and bending into a single process. By 1895, the EH Hotchkiss Company figured out how to feed a strip of staples into a machine. Thirty years later, the glued-together staples we know today were concocted by Jack Linksy (who would later launch the Swingline brand), making the device not only effective but also a time-saver.

The stapler would revolutionize a variety of fields, from printing to the surgical suite.

98 | ROOMBA

INVENTOR DR. HELEN GREINER | 2002

IF HUMANS CLEANED FLOORS the way robot vacuum cleaners do, they would be fired. Still, while Roomba and its offshoots may bump into walls and furniture like tipsy janitors, each generation of the robotic vacuum cleaner brings us closer and closer to *The Jetsons*.

Before the unveiling of the Roomba Floorvac in 2002, iRobot was focused on a very different purpose—building landmine-clearing robots that used the so-called crop circle algorithm to develop their travel patterns. Co-founder Dr. Helen Greiner adapted the same technology to enable her creation to sweep autonomously. Sensors indicate when a wall—or your cat—is about to be hit by the dirt-devouring disc.

Not yet in every household, the Roomba is the best-selling mobile robot in the world.

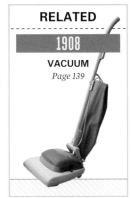

RELATED

1908

VACUUM
Page 139

97 AEROSOL SPRAY CAN

INVENTOR ERIK ROTHEIM | 1926

WHO'S RESPONSIBLE FOR THE GADGET that has enhanced everything from hair spray to cheese substitutes, making life easier for asthmatics, graffiti artists, and firefighters alike?

A Norwegian engineer, Erik Rotheim, held the first patent on aerosol technology. It got a boost when, in 1941, the USDA's Lyle Goodhue and William Sullivan used the newly discovered refrigerant Freon to enable the deployment of a lethal (to critters, anyway) mist by American troops fighting on insect-infested fronts. The "bug bomb" cocktail, held in a 16-ounce steel canister, consisted of Frecon-12, sesame oil, and pyrethrum (a natural insecticide derived from chrysanthemum blooms).

The plastic valve that tops most modern versions evolved from the work of Robert H. Abplanalp, whose replacement of the more expensive—and corrosive—metal valve made the carrier/sprayer less expensive and, thus, more popular with product manufacturers.

And according to the Aerosol Products Division of the Consumer Specialty Products Association, the majority of aerosol cans no longer use ozone-depleting CFCs.

96 QUICK-RELEASE SKI BINDINGS

INVENTOR HJALMAR HVAM | 1937

AFTER A WICKED 1937 ACCIDENT
on the slopes, made worse by fixed
bindings, Hjalmar Hvam had a vision: "When
I came out of the ether, I called the nurse
for a pencil and paper. I awakened with the
complete principle of a release toe iron."

It was uphill—or downhill, depending
on your perspective—from there. Hvam
was no novice on the slopes. A native of
Norway who had relocated to Oregon, he
was the first to ski off Mt. Hood's summit,
had placed first at the 1932 U.S. National
Nordic combined, and later coached the
1952 team at the Oslo Olympics. Hvam
developed his idea and steadily built a
following for his "Saf-Ski" bindings, which,
on impact, freed skis from boots—and
skiers' legs from knee-wrenching injury.

95 SUPER SOAKER
INVENTOR **LONNIE JOHNSON** | 1989

NASA scientist Lonnie Johnson has since had a hand in (arguably) more important work—including a thermo-electrochemical converter system—but it's the green and yellow blaster, and its many spin-offs and upgrades, that put him in the history books for millions of kids worldwide.

The aha moment came when Johnson was working on a heat pump and accidentally shot a stream of water across a bathroom where he was experi-menting. At first he manufactured the master blasters himself, before ultimately licensing the line to Hasbro.

Since 1991 no fewer than two dozen Super Soaker models have wrought backyard mayhem, but none is more coveted than the now-discontinued CPS 2000 Mk1. The most powerful water gun ever manufactured, it shot nearly a liter of water per second up to 50 feet, bringing new meaning to the warning "you'll shoot your eye out!"

Super Soaker has powered its water gun arsenal with pressurized reservoirs, spring power, electric motors, and most recently, a Constant Pressure System (CPS) with an expanding rubber bladder.

(94) | BLENDER
INVENTORS STEPHEN POPLAWKSI/ FRED OSIUS | 1919/1936

WANT TO KNOW THE HISTORY OF THE BLENDER? The story will depend on whether or not you want alcohol with that.

For the underage, let's use this one: In 1919 Stephen Poplawski crafted a mixer for use in restaurants. His device, according to his patent, pioneered "having an agitating element mounted in a base and adapted to be drivingly connected with the agitator in the cup when the cup was placed in the recess in the top of the base." You can trace Hamilton Beach appliances back to Poplawski.

For drinkers—or those wanting to use the device at home—the credit usually goes to Fred Osius who, with the financial help of bandleader Fred Waring, showed off his Waring Miracle Mixer at the 1936 National Restaurant Show in Chicago.

While it may seem to be all about the motor, the unheralded hero of blender development is its seal. The O-ring, patented in 1939, ensures that your drink ends up in a glass, not all over your kitchen.

93 | BRA
INVENTOR CARESSE CROSBY | 1914

▶ "I can't say the brassiere will ever take as great a place in history as the steamboat, but I did invent it," said Caresse Crosby, born Mary Phelps Jacob.

In 1914, she used handkerchiefs, ribbons, and cord to create an early version of the brassiere.

Well, maybe she did and maybe she didn't. "Invention" isn't a clear word in the history of the garment that shifted support of the breasts from the midsection, as with a corset, to the shoulders.

In ancient Rome, breasts were bound with cloth to prevent bounce during athletic events. But retail outlets didn't carry breast support items until 1902. The word brassiere didn't garner a mention in the American edition of *Vogue* until 1904.

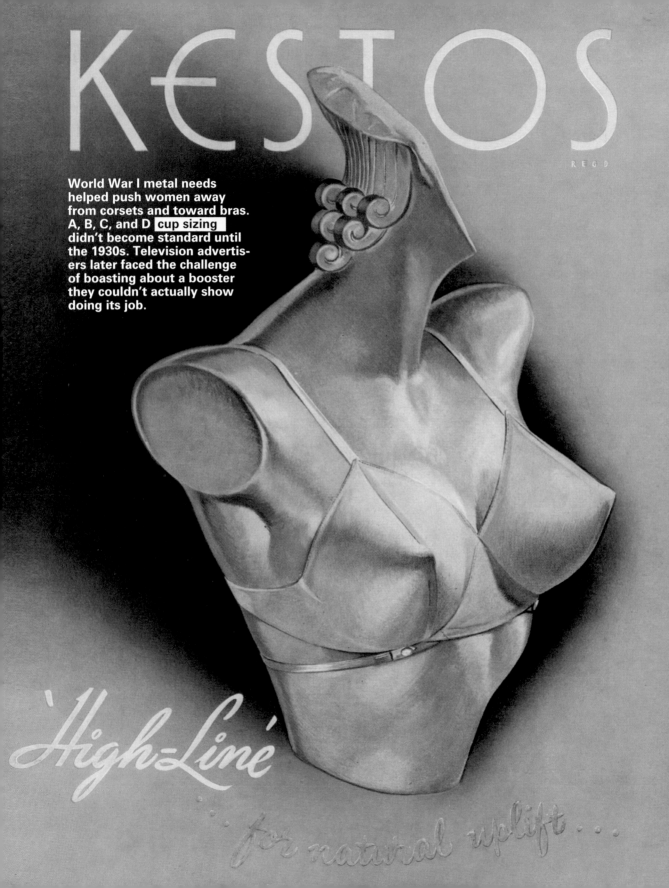

KESTOS

REGD

World War I metal needs helped push women away from corsets and toward bras. A, B, C, and D cup sizing didn't become standard until the 1930s. Television advertisers later faced the challenge of boasting about a booster they couldn't actually show doing its job.

'High-Line'

for natural uplift...

THAT TUB YOU USE TO CHILL BEER was patented as a "Portable Ice Chest for Storing Foods and the Like." Thank goodness it later got a shorter, cooler name.

While invented by Illinois resident Richard Laramy, W.C. Coleman popularized its use. According to company history, a Coleman engineer witnessed a kid blowing soap bubbles and applied its principles to portable "picnic chests," creating a protective, cool-keeping bubble with interior and exterior shells.

Igloo, founded in 1947, replaced wooden drinking buckets at job sites when it produced the first metal water coolers. Plastic became the material of choice in the '60s, and wheels were attached in 1994.

RELATED

1914

COLEMAN LANTERN
Page 85

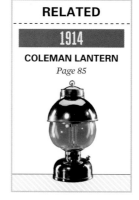

THE DIGITAL VIDEO RECORDING revolution, started in 1999 by TiVo (and its early competitor, ReplayTV), has been embraced by consumers who enjoy the flexibility the machines afford. And it's easy to understand why: DVR fundamentally changes the experience of watching TV.

First, it allows users to record programs to an internal hard drive for later viewing. Second, it lets viewers disrupt live TV—pausing, rewinding, and, perhaps most importantly, fast-forwarding delayed or recorded programming.

EVOLUTION OF THE RECORDER

1975	**1996**
VCR *Page 135*	**DVD PLAYER** *Page 76*

SINCE THE FIRST DISTINCTIVE CLICK OF A ZIPPO IN 1934, THE BRADFORD, PA COMPANY HAS PRODUCED NEARLY A BILLION LIGHTERS, ALL WITH A LIFETIME GUARANTEE.

Based on a design that's remained fundamentally unchanged from its original concept, the Zippo remained virtually unmatched for decades.

It began when former oil company exec George Blaisdell watched a friend light a cigarette with an ugly brass lighter from Austria. Blaisdell bought the U.S. rights, streamlined its brass case, chrome-plated the lid, refigured it rectangularly, and attached a spring-loaded hinge so the top could flick open. For a finishing touch, he surrounded the wick with a windscreen with a precise arrangement of holes that cut down the wind while allowing sufficient airflow.

The Virtual Zippo iPhone app exceeded more than 14 million downloads after its release in 2009 and is approved for concertgoers of all ages. All so you can keep that flame going during the band's encore of "Freebird."

A CLASSIC

Based on a design that's remained fundamentally unchanged from its original concept, the Zippo remained virtually unmatched for decades.

89 TEFLON PAN
INVENTOR ROY PLUNKETT | 1938

CHEMIST ROY PLUNKETT is responsible for one of the best known polymers in the world. With a high melting point of 620 degrees Fahrenheit and easy cleaning (pans can go in the dishwasher), the stuff DuPont called "the world's most slippery surface" has been separating eggs from pans for generations.

It started in 1938 as Plunkett experimented with Freon, a common refrigerant. When exploring its reaction to the colorless, odorless gas tetrafluoroethylene (TFE), Plunkett set aside a cylinder of TFE when it didn't discharge. He later returned to the cylinder and cut it open to see what had happened. What he found was a white power—the solidified gas—and, voilà, early Teflon was born.

First put to industrial uses, Teflon didn't earn the approval of the FDA until 1960.

88 | FLASH DRIVE
INVENTOR SHIMON SHMUELI | 1999

IN THE EARLY 1980S Toshiba engineer Fujio Masuoka developed flash memory, so named because the erasure process reminded a colleague of a camera flash. Intel went on to produce a commercial version in 1988.

But the good ship flash drive needed a way to dock. And although Intel's Universal Serial Bus (USB) provided part of the solution in 1996, data still didn't travel well until Shimon Shmueli, an IBM employee, invented the USB flash—or thumb—drive.

EVOLUTION OF DIGITAL STORAGE

1964	1971	1988
PORTABLE HARD DRIVE *Page 46*	**FLOPPY DISK** *Page 112*	**CD-ROM** *Page 94*

87 GINSU KNIFE

INVENTORS ED VALENTI AND BARRY BECHER | 1978

"WE LIKE TO SAY THAT GINSU MEANS 'I'll never have to work again,'" said marketing samurai Barry Becher in 2005. In the late 1970s, he and his partner Ed Valenti of Dial Media took a set of knives manufactured in Ohio, gave them a Japanese-sounding name, and changed the way America shops—and got rich doing it.

The memorable Ginsu TV commercial started with a tomato being karate chopped, followed by knife demonstrations on bread, meat, frozen packaged food, and even a tin can. But wait—

there's more! Every few seconds the offer grew from a single Ginsu knife to a whole set of utensils, including a carving fork, six-in-one kitchen tool, steak knives, and a spiral slicer.

Were Ginsu knives any better than the competition? Nobody really seemed to care. The hard-selling 1970s TV ads for these blades spawned *Saturday Night Live* spoofs, inspired Gallagher's comedy schtick, and ushered in the era of the infomercial. More than three million sets were sold before Warren Buffet purchased the company in 1985.

AS-SEEN-ON-TV

"I knew if we didn't capture the attention of the public in the first 10, 15 seconds of the commercial, we wouldn't be able to slide them into the rest of the ordering process," said Ed Valenti. One easy way to get attention: Try (and fail) to destroy your product.

▶ You heard that right. In the 1600s people actually used hornlike ear trumpets as hearing aids. (Beethoven was one of them.) It seemed like a great innovation after the previous not-so-sound technique: wooden hearing aids that were carved to look like the ears of keen-hearing animals.

The Akoulallion, the first American electric hearing aid released in 1898, was a tabletop model with a microphone and up to three pairs of earphones. Mass-produced, in-ear versions didn't arrive until 1966, followed by digital versions in the late 1980s.

Although they can be as powerful as a PC, modern **hearing aids still contain four basic elements: a microphone, an amplifier, a speaker, and a battery.** Instead of simply acting as amplifiers, they're also selective hearing devices. They reduce wind noise, raise quieter sounds, lower louder ones, and control feedback.

Whatever the technology, there's still a problem that can't be solved by mechanical means: According to the National Institute of Health, only one out of five people who could benefit from a hearing aid actually wears one. Denial, it seems, is the industry's biggest obstacle.

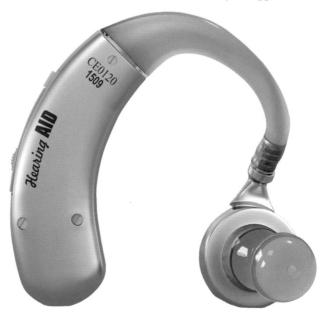

85

SUNGLASSES
INVENTOR SAM FOSTER | 1929

WHO'S REALLY BEHIND THOSE FOSTER GRANTS?

Probably the Inuit people, who fashioned snow goggles some 2,000 years ago by cutting slits into bone, ivory, or wood and using caribou sinew as a strap. They rubbed gunpowder or soot on the outside to reduce glare.

But shades as we know them today originated in 1929 on the Atlantic City boardwalk, where Sam Foster, founder of Foster Grant, first hawked mass-produced plastic sunglasses to beachgoers. Edward Land polarized the lenses six years later.

General Douglas MacArthur and Tom Cruise in Top Gun can trace their iconic Ray-Ban Aviators back to 1937 when Bausch & Lomb added a green tint to metal-framed sunglasses to reduce glare for Army pilots.

RELATED

1870

OUTBOARD MOTOR
Page 96

Coffee has greatly impacted the world, not just for its value as a commodity, but also for its ability to wake up paupers, kings, and everyone in between.

Coffee history can be divided into epochs. There was the early era where history is fuzzy, myth is prevalent, and the Turks famously brewed. Then in 1818, the percolator period began, going electric in the latter half of the century.

Mr. Coffee deserves its own period, launching automatic drip brewing in the home in 1972 and adding filters three years later. The much-copied

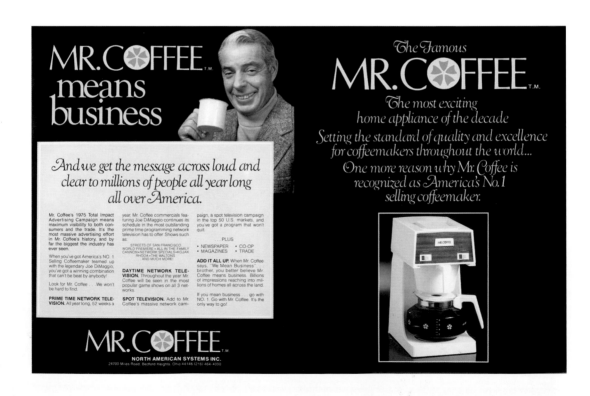

process involves the heating of water in small tubes. When the water boils, it rises in the tubes and passes through the coffee grounds. After picking up the oils in the coffee grounds, it drips into the waiting pot below.

(83) TOASTER

INVENTOR CHARLES STRITE | 1919

YOU SIMPLY PUSH DOWN THE HANDLE, and in a few moments the bread pops up all by itself, freshly toasted.

What happens between the push and the pop? Radiant heat raises the bread's outer temperature to about 310 degrees Fahrenheit, when something called the Maillard reaction caramelizes the sugars and starches.

Scorching bread—for taste and preservation—is nothing new. In pre-electric days, it was done using a fireplace tool that held the bread over a flame. But with the death of the fireplace came the birth of the electric pop-up toaster, invented by Charles Strite. With a little financial help from his friends, he sold 150 of them to a restaurant chain after filing for a patent in 1919. By 1926 he brought the Toastmaster into homes, giving rise to that wonder of modern food science, the Pop-Tart, nearly 40 years later.

COMEDIANS HAVE DESCRIBED flashlights as places to store dead batteries. And there's a grain of truth to that, as many have discovered during a blackout.

But while the flashlight may seem like a simple device, its origins aren't exactly a straight beam of light. The invention of the dry cell battery (specifically, the relatively lightweight D cell) and the lightbulb in the 1880s paved the way for the flashlight. The first released in 1898 became a much safer alternative to kerosene lanterns and candles.

Some early versions didn't have an on/off switch. They were named "flash lights" not because they could quickly turn on but because of their inability to provide a steady beam.

While others contributed to design and technical improvements, the American Electrical Novelty & Manufacturing Company became the great popularizer of the flashlight. For its first catalogue, the company distributed flashlights to police officers and picked up their endorsements, which it printed. Public demand grew, and soon the name of the company changed, too, lifting inspiration from one of its marketing slogans and shortening "Ever ready" to Eveready.

COPYRIGHTED BY THE AMERICAN ELECTRICAL NOVELTY & MFG. CO., N.Y.

The first flashlight consisted of a paper tube, a brass reflector and end caps, and a bulls-eye glass lens. The most antiquated part of all: The three D batteries were included.

81 LEAF BLOWER

INVENTOR **VARIOUS** | 1950s

n the 1950s hedge-trimmer manufacturer Echo, Inc. discovered that consumers often turned its engine-powered duster/mist blower, which gardeners wore on their backs, into a simple but powerful air blower. Former jet engine technician H.L. Diehl made things official in 1959 by developing a "walk-behind lawn vacuum and leaf blower." Then 1969's Giant-Vac changed the game even more by introducing a blower that also collected the leaves for easy disposal.

The biggest downside: noise and pollution. But that didn't stop the spread of leaf blowers in suburbs nationwide, exponentially reducing yard maintenance time.

A combination of civic laws (Los Angeles banned gas blowers within 500 feet of residences) helped push manufacturers toward plug-in versions and later, cordless models using NiCad batteries.

All of which allowed for the 1998 creation of one of Canada's more interesting sports: leaf blower hockey, whose inaugural game pitted the Fallen Leaves against the Windbreakers. Try doing *that* with a rake.

RELATED

1830

LAWNMOWER
Page 104

You've got leaves
on your lawn. You
want to move/remove
them. You don't
feel like raking. The
solution: **air power**.

INVENTOR GEORGE SNYDER | 1820

CHINESE PAINTINGS SHOW FISHING RODS AS FAR BACK AS 1195. AND THE BRITISH WERE EXPERIMENTING WITH BETTER WAYS TO LAND FISH IN THE 1600s BY USING A "WIND" (RHYMES WITH KIND)—A PULLEY SYSTEM FOR STORING EXCESS LINE.

But it wasn't until the 1800s, when Kentuckian George Snyder designed his fly cast reels, that these handy sportsmen's items began to resemble the ones pulling fish in today.

Backlash proved an ongoing problem. So were the birds' nests of tangled lines. The breakthrough for novice fisherfolk came after World War II when the Zero Hour Bomb Company's patent on an oilfield time bomb approached expiration. When **R.D. Hull presented the company with plans for an easy-to-use enclosed-spool fishing reel,** Zero Hour rebranded itself as Zebcom—and sold thousands of reels by 1956.

Winding and unwinding were handled with the simple push of a button, stopping the line when the lure hit the water. Soon more fish were caught with spinning and spincast reels than by any other casting method.

When fishermen claimed they caught a fish *this* big, they might have actually been telling the truth.

RELATED

1940

FIBERGLASS FISHING ROD
Page 54

how the zero hour bomb company made tulsa the fishing tackle capital of the world

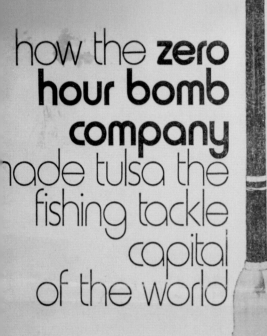

[I]n a remote corner in an oil well bomb factory has come [the] most famous name in fishing tackle. A name that [mea]ns quality, economy and fun. A name that has [mad]e Tulsa as famous for fishing as oil. The [nam]e: ZEBCO!

[It] started back in 1949 when R. D. Hull [work]ed at the Zero Hour Bomb Company [with] a new kind of fishing reel. It resembled [a b]eer can with a hole in one end. He [nee]ded a place to produce this new reel. [Ze]bco (contraction of Zero Hour Bomb [Com]pany) agreed to give him a corner of [the p]lant to do it in. Soon attentions were [drawn] to the commotion in that back corner, [and] before long, Zebco's main concern was [fishin]g reel manufacture, with bomb produc-[tion a]s a sideline.

[From] the very beginning, Zebco did things dif-[feren]tly. Never before had a reel maker installed [mono]filament line on every reel sold. It was unheard [of fo]r a reel company to start production of a whole line [of p]erfectly balanced rods for use with specific reels. And, [who'd] dare stick their neck out with marketing innovations [like in]cluded packaging a reel and rod together and selling them [as a] single unit? Or meet the challenge of competitive, low-price [impo]rt reels with a better quality economy reel of their own? Zebco did [all t]hese things and more. Things that were different, but things that [mad]e sense.

[Zeb]co has made great progress in 21 years. The small workshop, where a [staff] of four people turned out 53 reels a day, has grown into a 16 acre [facili]ty with over 500 employees who assemble more than 20,000 reels [ever]y day. Suppliers of Zebco parts in the Tulsa area alone employ at least [500] people. So you might say the Zero Hour Bomb Company is still ex-[plod]ing . . . but in terms of manpower and jobs instead of nitro.

[Doin]g different made Zebco the number one name in the fishing tackle indus-[try. I]t has made Tulsa, Oklahoma the Tackle Capital of the World. And it has [mad]e the fun of fishing easy for millions of men, women and children around [the g]lobe. Zebco keeps on doing things differently from other tackle manufac-[turer]s . . . things like developing a new line of outdoor recreational equipment [like p]ropane-fueled catalytic heaters and lights. And making more plans for ex-[pand]ing the line to include stoves and other camping equipment.

[ZEB]CO does things differently. And it's paying off for Oklahoma. Paying off with [more] jobs, more capital investments and more dollars on the Oklahoma stringer.

Consumer Division Brunswick Corporation
P. O. Box 270, Tulsa, Oklahoma 74101

ZEBCO®

For truly *revolutionary* SPINNING and CASTING make it a **ZEBCO REEL**

All reels specially gift wrapped for holiday giving.

MODEL 33
$19.50
with 150 yards of 6-lb. test mono-filament line in-stalled*

new! ZEBCO Spinning Reel

Designed to the special demands of Amer-ican fishermen the increasingly popular Zebco Spinning Reel is truly revolutionary. No longer do you need an "educated" finger to control your cast.

Instead, you merely depress the simple thumb control, then release it as the rod tip moves forward, and the lure (as light as you care to use) is on its way in the *easiest* cast you ever made. The first turn of the crank picks up the line to start a liquid-smooth retrieve. Hook a big one? A couple of turns on the handy drag button tightens the line to slow down your fish and keep him fighting all the way!

Another Zebco advantage is that the Model 33 is right at home on a straight-handle spinning rod, though it was primarily designed for the more com-fortable handle of the off-set type.

Caught with a ZEBCO!

Harry A. Norris. Jr., of Bay City, Texas, recently set a new world record with this 70-lb. cobia, caught on a 10-lb. test line with a Model 11 Zebco Casting Reel off the mouth of the Colorado River in the Gulf of Mexico.

new! ZEBCO Casting Reel

The improved version of the original standard Zebco, the reel which made backlash a thing of the past.

● Stainless steel covers
● Interchangeable spool
● Handy Zebco thumb control
● Improved spinner head for lighter lures

MODEL 11
Now $12.50
with 85 yards of 10-lb. test monofilament line installed*

The expert's choice for performance, the beginner's for fishing ease.

Super ZEBCO Casting Reel

A performance-proved favorite which in just two years has become one of America's most popular casting reels!

● Improved adjustable drag
● Liquid-smooth retrieve
● Streamlined stainless steel covers
● Handy Zebco thumb control
● Takes lures as light as ¼ oz.

MODEL 22
$17.50
with 100 yards of 8-lb. test monofilament line installed.

*Extra spool without line $.75; with 4, 6, 8 or 10-lb. test monofilament line installed $2.00.

Ask Your Dealer For A Demonstration

Manufactured by ZEBCO COMPANY, Tulsa, Oklahoma

Swiss cutler Karl Eisener invented the ubiquitous multipurpose tool because he was bothered by the fact that Swiss soldiers carried German-made knives. He formed the Association of Swiss Cutlers to correct that with the "Soldatenmesser." How did it become the knife we all know? By World War II U.S. soldiers who couldn't pronounce the Swiss name were just calling it the Swiss Army Knife.

The original wasn't quite as packed with sub-gadgets as later models—it featured just four tools: a blade, an awl, a can opener, and a screwdriver. The 1897 edition, nicknamed "Schweizer Offiziersmesser," kicked in a shorter blade and a corkscrew and, perhaps more importantly, modified the look to include the now iconic red handle.

The Swiss Cross became part of the design in 1909, and the newly invented stainless steel was adopted as the material of choice in 1921. And it keeps on evolving. The "SwissChamp xlt" now has 35 tools, including a wood chisel. At this rate, all of this book's gadgets may fit on it.

RELATED

1983

LEATHERMAN
Page 108

78 CAN OPENER

INVENTOR EZRA WARNER | 1858

BEFORE EZRA WARNER INVENTED the can opener in 1858, canned food had to be carefully chiseled out with a hammer. Warner's bayonet and sickle combination punctured lids without leaving a dangerously sharp edge, but it still wasn't a household item. Most shoppers got theirs opened at the grocery store.

A hundred years later—and a mere year before the invention of the pop-top can—G.E. rolled out the automatic electric can opener. Manufacturers liked it because the tin used in cans could be made thinner and easier to puncture, thus cutting costs. Consumers liked it as a wrist-saver.

There were innovations in between. In 1870 William Lyman patented a design that punctured the center of the can with the gadget serving as a pivot to allow a cutting wheel to make its trek. In 1925 the Star Can Opener Company flipped the wheel blade so that it cut under the can rim.

77 DVD PLAYER
INVENTOR **VARIOUS** | 1996

When videotapes gave way to DVDs, the industry largely welcomed the change. Not only was quality significantly greater, but so was life span, maneuverability, and storability. And you could play your CDs on these players as well.

EVOLUTION OF THE RECORDER

1975	1999
VCR	**DVR**
Page 135	*Page 62*

The only problem for customers was having to purchase new versions of all of those tapes filling their cabinets. The manufacturers didn't mind this, of course.

The prototype for the DVD player, developed in 1994 by Toshiba, was a circuit board stack nicknamed "the fire watchtower." Though unstable, it proved that DVD quality crushed that of VHS. Players came out in 1996, and *Twister*, the first Hollywood film pressed to DVD, was released in 1997. Soon commentary and other special features impossible on VHS became commonplace. (Film critic Roger Ebert recorded so many commentary tracks that, when he lost his voice after cancer surgeries, he used these tracks to build his computerized voice.) The knockout blow delivered to VHS by DVDs came in 2005, when Walmart stopped carrying the now-clunky older format.

76 CHAIN SAW

INVENTOR EMIL LERP | 1927

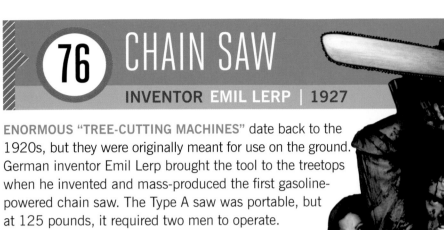

ENORMOUS "TREE-CUTTING MACHINES" date back to the 1920s, but they were originally meant for use on the ground. German inventor Emil Lerp brought the tool to the treetops when he invented and mass-produced the first gasoline-powered chain saw. The Type A saw was portable, but at 125 pounds, it required two men to operate.

Thirty years later, improvements in aluminum and motor design allowed solo backyard warriors to use solo saws to prune trees and cut firewood.

The chain saw also became part of pop culture with a grisly star turn in the 1974 release of *The Texas Chainsaw Massacre* (original working title: *Headcheese and Leatherface*).

75 ELECTRIC BLANKET

INVENTOR SIDNEY RUSSELL | 1912

THERE'S A FINE LINE between a giant heating pad and an electric blanket, but wherever you draw that line, American physician Sidney I. Russell usually gets the credit.

Then breakfast cereal inventor and Battle Creek Sanitarium director John Kellogg, who advocated sleeping outdoors to promote general wellness, picked up on the idea. His "thermo-electric" blanket enabled residents to enjoy fresh night air regardless of the season.

Think that's odd? In 1967 Vince Lombardi, general manager of the Green Bay Packers, worked out a deal with General Electric to install giant, electric-blanket–like coils under the turf at Lambeau Field to counter the venue's reputation as The Frozen Tundra.

No matter the size, the concept is simple: A current runs through resistive wire wound through the blanket. The tricky part is figuring out how to keep a human toaster safe.

Diagram of the
New Improved
GILLETTE

OVERHANGING CAP
ADJUSTABLE
SHORT FLEXURE
CHANNELED
GUARD
FULCRUM
SHOULDER
DIAMOND
KNURLING
PAT. JAN 13, 1920

THAT DELIGHTFUL FEELING of freshness after a *comfortable* shave is enjoyed by the multitude of New Improved Gillette users — men critical of razor service who demand the utmost in shaving results.

The price is $5 and up

"Three Reasons" is the title of a convincing booklet sent upon request

The New Improved

Gillette

SAFETY — Gillette — RAZOR

GILLETTE SAFETY RAZOR CO., BOSTON, U. S. A.

IN THE DAYS OF STRAIGHT RAZOR SHAVING,
a single nick could easily become a major bloodletting. Safety razors existed, but their blades had to be sharpened frequently. Then King Camp Gillette had an idea: a safety razor with disposable steel blades for easy replacement. The idea should've been a welcome innovation in men's grooming, but it wasn't immediately appreciated. It took Gillette six years to find an engineer willing to give it a shot—a man appropriately named William Emery Nickerson. Together they launched the American Safety Razor Company, later renamed Gillette, and changed the landscape of shaving…and men's faces everywhere.

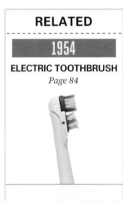

RELATED

1954

ELECTRIC TOOTHBRUSH
Page 84

◄

While other safety razors used forged steel blades, Gillette's were the first to use stainless razor steel. The inventor appreciated a close shave as much as anyone, but he was most interested in the profits that came with selling an item that was used a few times and then discarded.

73 PRINTER

INVENTOR XEROX | 1968

PRINTERS GO AS FAR BACK as the 14th century, when Johannes Gutenberg's movable printing press first started printing bibles. But it wasn't until the 1950s, when computer granddaddy Univac had its own Remington-Rand machine, that files could be printed from a computer. And in 1968 Seiko marketed a tiny printer that went with a calculator.

The real breakthrough came in the late 1960s when Xerox's Gary Starkweather combined its copier tech with a laser beam—an idea that Starkweather later said his company didn't have a strong interest in developing. Wiser heads prevailed, and the 9700 Electronic Printing System, the first laser printer (which oddly predates the dot matrix printer), went to work.

72 STOPWATCH

INVENTOR TAG HEUER | 1916

ACCORDING TO the National Institute of Standards and Technology, a second is "the duration of 9,192,631,770 periods of the radiation corresponding to the transition between two hyperfine levels of the ground state of the cesium-133 atom."

As difficult as that mouthful is to rattle off quickly, it was even more difficult to measure, until the invention of the stopwatch.

When the TAG Heuer Mikrograph stopwatch arrived in 1916, it allowed the measurement of time with unprecedented accuracy—down to 0.01 second. It was no surprise that it became the first official stopwatch of the post-war Olympic games.

The next decimal place leap didn't come until 1971, when measurements to 0.001 second—and a whole new blink-of-an-eye set of records—became possible. When Michael Phelps beat Milorad Cavic at the 2009 Summer Olympics, the tape had to be reviewed 0.0001 second at a time to determine the winner. Talk about hyperfine!

KODAK CAROUSEL

INVENTOR LOUIS MISURACA | 1962

AD MAN DON DRAPER SAID IT BEST IN HIS PITCH TO KODAK EXECS IN AN EPISODE OF *MAD MEN*: "THIS DEVICE ISN'T A SPACESHIP, IT'S A TIME MACHINE."

The Carousel wasn't the first slide projector. But its innovations to the gadget allowed for simple drop-in loading, catering to fumble-fingered folks anxious to show slides of their European vacation to anyone willing to watch. The Carousel held a then-remarkable 80 slides, and allowed for continuous play when the standard rectangular slide tray was switched out with a circular one.

These days, it's difficult to imagine projecting your recent history onto a wall . . . unless it's on Facebook. By 2004 interest in old-school slideshows had waned so much that Kodak dropped the product. But in the 1960s and '70s—when the bulk of the 35 million projectors were sold—the common question at the photo processing lab was "Prints or slides?" No self-respecting graduation party went on without a slideshow, and closet space had to be allocated to boxes of watched-once memories.

70 BOOMBOX

INVENTOR ROYAL PHILIPS ELECTRONICS | 1969

WHAT WAS SO IMPORTANT, CULTURALLY, ABOUT AN OVERGROWN TAPE PLAYER?
In the pre-Internet days of the late 1970s and early '80s, the boombox allowed anyone unsatisfied with commercial radio to play their own tunes, broadcasting uncensored music to everyone within earshot.

The derisively labeled "ghetto blasters" could also record off the radio, or, thanks to double cassette capability, from other tapes. With mixtapes making the rounds, a radio station was no longer needed for a tune to go viral.

Technology soon allowed for powerful sound systems in smaller and smaller packages. But unlike the headphoned Walkman, the boombox took the party to the streets, channeling hi-fi audio tracks into hip-hop rebellion. It defined the era—and it's tough to beat the image of the boombox, accompanied by b-boys on a refrigerator box dance floor.

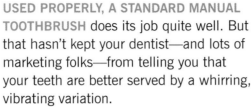

USED PROPERLY, A STANDARD MANUAL TOOTHBRUSH does its job quite well. But that hasn't kept your dentist—and lots of marketing folks—from telling you that your teeth are better served by a whirring, vibrating variation.

Swiss Dentist Dr. Philippe Guy Woog gets credit for the first electric toothbrush. His creation, the Broxodent, was originally targeted to those who lacked the motor skills for standard brushing. Powered brushes also proved useful in the late 1950s when U.S. Navy submariners, whose diets subsisted largely of mushy canned food for months on end, needed something to give their gums some stimulation. The gadget later found a wider audience.

Tweaking continues with the recent fusing of dental cleaning and Bluetooth technology. The Oral-B Triumph with Smartguide not only keeps track of our brushing time, but also alerts you, graphically, when you're brushing too hard.

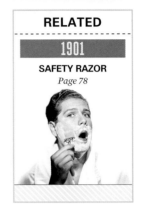

RELATED

1901

SAFETY RAZOR
Page 78

PULLING THE PLUG

The Broxodent plugged right into the wall and ran on AC line voltage, which could be hazardous in a wet environment. In the early 1960s, General Electric introduced the cordless Automatic Toothbrush with rechargeable NiCad batteries.

IN THE LATE 1800S AND EARLY 1900S, RURAL PARTS OF THE COUNTRY REMAINED LARGELY OFF THE GRID AND UNELECTRIFIED. THAT MEANT THE WORKDAY ENDED WHEN THE SUN WENT DOWN.

Then along came Coleman, a typewriter salesman who revolutionized the illumination business with his Quick-Lite lantern. By 1920 his company had sold 50,000 of the devices.

As electricity spread to rural areas, Coleman rebranded his lanterns as an outdoorsman's essential. Coleman lanterns lit up the woods when the popularity of camping skyrocketed after World War II. No wonder the *Saturday Evening Post* wrote, "Except for Thomas A. Edison, Mr. Coleman may be responsible for the creation of more bright light than any other man."

RELATED

1951

PICNIC COOLER
Page 62

BINOCULARS

INVENTOR UNKNOWN | 1700s

▶ "Universal alone now makes more binoculars for the Army, Navy and Marines, and the United Nations, than the entire industry made before the war," boasted Universal Camera Corporation during World War II.

The company was nearly wiped out when the war took out most of its European film markets, but it rallied thanks to rethinking itself as a manufacturer of binoculars.

The device itself was nothing new. Gadgets that used side-by-side telescopes, one for each eye, had been experimented with since the 17th century. But because of mass production and wider distribution, post-war amateur ornithologists flocked to fields and forests armed with affordable binoculars. Baseball fans and concertgoers got up-close-and-personal views from the cheap seats. And the amorous got better about closing their windows.

These days, binoculars aren't just for bringing objects closer visually. Some optical communication models now come equipped with USB and Ethernet adapters and receivers that are capable of beaming secure voice and video information using eye-safe infrared LEDs.

Even more reason to make sure those blinds are closed.

TAPE MEASURE

66

INVENTOR JAMES CHESTERMAN | 1821

FIRST THERE WAS THE YARDSTICK; then came the ribbon tape measure. But major breakthroughs in tape measure design came later—first, when Londoner James Chesterman found that excess coils of flat metal tape could be marked and used for measuring; and second, when Alvin J. Fellows of New Haven, CT was granted a patent for a similar device with one key innovation: His could be locked in place, if desired, thanks to a spring clip.

This new device, housed in its own case, didn't become a toolbox staple until after World War II. Carpenters quickly caught on to how it sped up construction as they worked to meet the demands of a major building boom: 14.1 million homes in a decade.

Now you can find tape measures with ink tips to mark without a pencil and others with LCD readings. None, however, comes equipped with a helmet to keep you from being smacked in the head by the rapidly receding metal tape.

ZIPPER

65

INVENTOR GIDEON SUNDBACK | 1914

IT WAS CLEAR from the beginning that a lightbulb would change the world. The same can't be said about the zipper, which was used on mailbags for years before the technology was transferred to shoes and then to just about every other article of clothing.

Initially known as a Clasp Locker or Unlocker for Shoes, the 1891 version designed by Whitcomb Judson had the basics of what we know as a zipper today: a fastener guide paired with clasps, albeit ones that didn't quite work.

In 1911 Swedish immigrant Gideon Sundback, inspired by Judson's designs, began making adjustments. He upped the number of fastening elements from four per inch to 10 or 11 per inch and earned a 1917 patent for his "separable fastener."

64 DERRINGER

INVENTOR HENRY DERINGER | 1825

▶ DERRINGER BECAME A NOTORIOUS PART OF AMERICAN HISTORY ON THAT FATEFUL 1865 NIGHT WHEN PRESIDENT ABRAHAM LINCOLN ATTENDED *OUR AMERICAN COUSIN* AT FORD'S THEATER IN WASHINGTON, D.C.

The one-shot concealed weapon, designed by Henry Deringer (with one "r" instead of two) in 1825, was certainly not crafted for political assassination. But it was carefully created to be effective at close range. At his plant in Philadelphia, Deringer, the son of a riflemaker, first manufactured muzzle-loaders with flintlock firing, later making

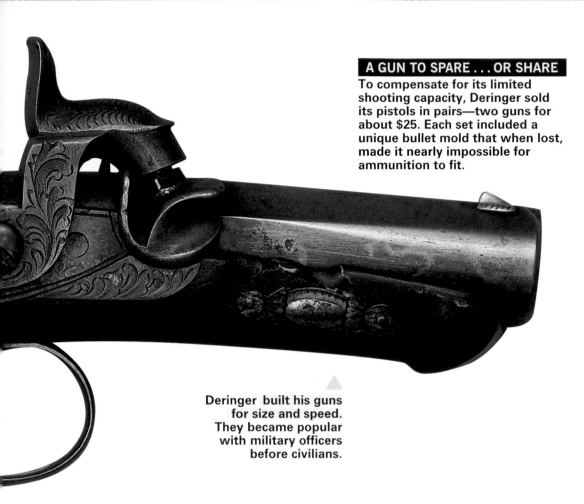

Deringer built his guns for size and speed. They became popular with military officers before civilians.

use of percussive cap technology. Instead of focusing on premium accuracy, he built his guns for size and speed.

Imitators came along almost as soon as Deringer's one-shooter hit the market in 1825, and Derringer quickly became a widely used generic term for a pistol. Not sure of the authenticity of yours? The genuine gun has the words "Deringer Philadelphia" engraved on the action and cannot have been manufactured after 1870, when Deringer ceased production.

THE LUNCHBOX ITSELF ISN'T A NEW CONCEPT. LONG BEFORE THE BRANDED BOXES WE THINK OF TODAY, WORKERS IN INDIA CARRIED MEALS IN STACKING TIFFAN CANISTERS, AND THE JAPANESE PACKED WOODEN LACQUERED BENTO BOXES.

These—and American lunch kits—were targeted to grown-ups. Then Mickey Mouse got his mug slapped on one in 1935, and the rules changed. And when Aladdin Industries launched the Hopalong Cassidy kit in 1950, it kicked off the lunchbox boom: Between 1950 and 1970, 120 million lunchboxes shot off store shelves as America's youth scrambled to get the latest in branded boxes. Accompanied, of course, by a keep-the-cold-cold/keep-the-hot-hot thermos with a vacuum gap between the outer and inner surfaces.

RELATED

1946

TUPPERWARE
Page 47

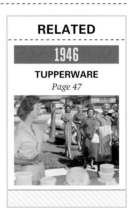

LUNCHTIME STATUS SYMBOLS

Branding with trendy characters not only helped popularize the lunchbox, it also ensured new sales each year as characters and themes became passé and having a new lunchbox became part of the transition into a new school grade.

Steve Canyon

HATS OFF TO MOM FOR

Gunsmoke

HER WHOLESOME LUNCHES

Globe Trotter

AND THE WONDERFUL KITS

Junior Miss

(AND KIDS) SHE PUTS 'EM IN!

To fun-loving small-fry, what's *ON* their lunch kits is just as important as what's *IN* 'em . . . and Aladdin goes all out to give them their up-to-the-minute favorite TV and Story book idols—in vivid, non-fading color on both the roomy steel kit and matching ½-pint performance-guaranteed Aladdin Vacuum Bottle. Kids love 'em.

Guaranteed by **Good Housekeeping**
REPLACEMENT OR A REFUND OF MONEY IF NOT AS ADVERTISED THEREIN

USE-TESTED by McCall's

. . . them, buy them . . . at your nearby hardware, drug, variety, department store or supermarket

Aladdin®

SCHOOL LUNCH KITS AND MATCHING ½-PINT VACUUM BOTTLES

62 CHARCOAL GRILL

INVENTOR GEORGE STEPHEN | 1951

BEFORE ENTERING DESIGN HISTORY for its signature kettle grill, Weber Brothers Metal Works was an Illinois outfit that produced among other things, buoys. In 1951 worker George Stephen borrowed the same shape and material used in a harbor guide and built the first Weber kettle. Stephen started selling it through the barbecue division he formed and later bought the entire factory.

The kettle proved a hit, particularly in the Brat Belt of Midwestern suburbs, where backyard chefs driving the barbecue boom embraced its cooking prowess. Today, even after gas grills have made barbecuing easier, the Weber charcoal grill remains a backyard staple.

61 SMOKE DETECTOR

INVENTORS RANDOLPH SMITH, KENNETH HOUSE | 1960

FEW GADGETS IN HISTORY have had the primary and almost exclusive purpose of saving lives. The fact that the basic technology that made the smoke detector possible was discovered accidentally is staggering, especially when you consider that it has cut fire deaths in half since the 1970s.

Duane Pearsall had merely wanted to curb static in photo darkrooms, but noticed that the meter measuring ion concentration on his static-control device would flat-line whenever cigarette smoke hit it. "By accident, we had discovered how to make an ionization smoke detector," Pearsall said.

In 1969 Randolph Smith and Kenneth House patented a battery-powered smoke detector for home-use based on this early discovery. Current models rely on perhaps the cheapest nuclear technology you can own: a chunk of Americium-241. When smoke particles attach to and neutralize its charged ions, the electric current falls and sets off the alarm.

MOOG SYNTHESIZER

INVENTOR ROBERT MOOG | 1965

THE MOOG SYNTHESIZER might have earned its spot in this book simply because it made the Doors' *Strange Days*, Simon & Garfunkle's *Bookends*, and the Beatles' *Abbey Road* possible. But the influence of the instrument goes beyond those key moments in popular music history.

Among the first widely used electronic instruments, the Moog used analog circuits to generate sound electronically. Robert Moog formed the R.A. Moog Company after building theremins, early electronic instruments often used in creepy horror movie music—and the Beach Boys hit "Good Vibrations."

Moog crafted what he intended to be an affordable (and smaller) synthesizer. But the device didn't just generate better sounds than other synthesizers. It could be controlled by a keyboard instead of punch cards, making it an instrument, not just a machine. By the early '70s, the Moog was part of the music-producing landscape. "Switched-On Bach," entirely crafted on the Moog, topped the charts. The soundtrack of *A Clockwork Orange* was Moog-dominated. Keith Emerson of Emerson, Lake & Palmer carted a 10-foot-tall version to concerts, and George Harrison had one in his home.

The acceptance of electronic music proved a crucial step in developing audio technology for computers, mobile phones, and stereos. And it sounded pretty cool, too.

WELL INTO THE LATE 1980s, computer games and other programs required multiple floppy disks in order to operate. Each of these storage devices could handle about 1.5 megabytes of information, hardly enough to get most serious games going, let alone cool animation. *Alone in the Dark*, the first 3D survival horror game, took four such disks.

Then along came the CD-ROM—short for compact disc read-only memory—which could package most programs on a single disc, making it much easier to disappear into *Myst*. These circular wonders had more storage capacity than most early computer hard drives. Encyclopedias were made available on CDs—allowing their contents to be more easily cribbed by students doing research papers.

The hardware changed to accommodate. Most 1990s computers came with a CD drive. Games became more interactive because of the speed by which the discs could connect to information. The creation also led to the demise of the VCR, effectively turning computers into movie players. And computers themselves were suddenly seen by the average consumer as essential, useful, and entertaining—a legacy that continued well after the discs found their way to garage sales and then to dumpsters.

EVOLUTION OF DIGITAL STORAGE

1971	1964	1999
FLOPPY DISK	**PORTABLE HARD DRIVE**	**FLASH DRIVE**
Page 112	*Page 46*	*Page 64*

DIGITAL STORAGE IN LEAPS AND BOUNDS

WHEN IT COMES TO DIGITAL STORAGE, SIZE CUTS BOTH WAYS—EVER SMALLER DEVICES WITH EVER GREATER CAPACITY. HERE, 40 YEARS OF GROWTH (AND SHRINKAGE).

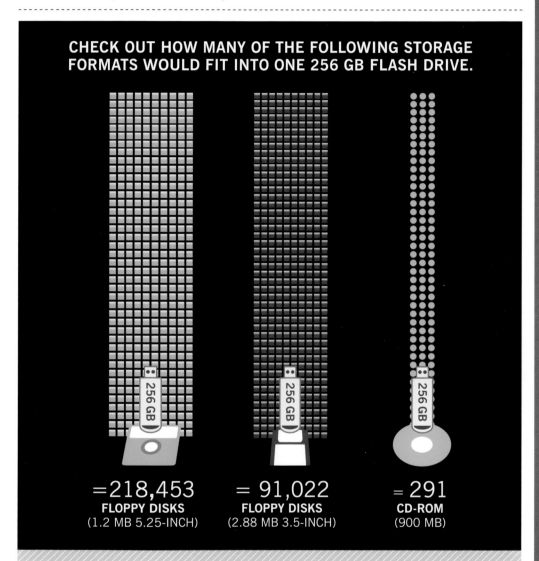

CHECK OUT HOW MANY OF THE FOLLOWING STORAGE FORMATS WOULD FIT INTO ONE 256 GB FLASH DRIVE.

256 GB

=218,453
FLOPPY DISKS
(1.2 MB 5.25-INCH)

= 91,022
FLOPPY DISKS
(2.88 MB 3.5-INCH)

= 291
CD-ROM
(900 MB)

In an analogue to the biyearly leaps forward in processing power described by Moore's law, storage capacity per dollar has also increased exponentially (the storage devices are shown here at their maximum capacities). Consider: In 1978 a 5.25-inch dual-density floppy disk that held 360 kilobytes cost about $12. Today, that same $12 buys an 8-gigabyte flash drive with more than 20,000 times the capacity.

THE STORY GOES THAT NORWEGIAN IMMIGRANT OLE EVINRUDE was inspired to invent the outboard motor in 1911 when he crossed a lake to get ice cream during a picnic. By the time he returned, the ice cream had melted, and Ole vowed to pick up the pace of water crossing and suffer no more.

RELATED

1929

SUNGLASSES
Page 67

Evinrude wasn't the first person with the idea, though. Cameron Waterman, an appropriately-named Yale engineering student, actually beat him by three years, patenting a detachable "boat-propelling device" in 1907.

But even before him, there was Gustave Trouvé, who, in addition to developing an early telephone, electric jewels, and a gunpowder ornithopter, took a motor that worked on land and attached it to a boat.

The modern outboard motor didn't become an important American gadget until the '50s. In 1956, 600,000 new outboards hit the water, and the country's thirst for powered-up aquatic good times began in earnest. Today Evinrude's ice cream would stay frozen. His beer wouldn't even get warm.

HAIR DRYER

INVENTOR ALEXANDRE GODEFROY | 1890

BEFORE THE 1890s, DRYING HAIR WITH A MACHINE WAS UNHEARD OF.

That changed when French hairstylist Alexandre Godefroy redirected hot air from a gas stove chimney onto the heads of his clients. The hooded metal bonnet ran on a power cord and hand crank. Très chic and très dangerous.

Soon tinkerers were figuring out how to use their vacuum cleaners and toasters to do the job.

Handheld home models didn't arrive until the 1920s, but the two-pound gadgets were more likely to overheat than actually dry hair. Plastics, developed in the 1950s, made the machines lighter. Although the technology—fan-blown air over heated plates—has stayed consistent, laws passed in the '90s made hair dryers significantly safer.

Today most homes—and 96 percent of hotel rooms—have a hair dryer to make grooming easier. The hairstyling tool also comes in handy when removing bottle labels and heating up cold diesel engines.

HAIR APPARENT

Today's hair dyers include a number of attachments, all rooted in science. Wide diffuser attachments keep air from blowing around hair while it dries, while airflow concentrators do the opposite and speed up drying. Ionic hair dryers release negatively-charged ions to reduce static electricity and enhance shine. Even the most basic dryers have a cool button for setting your do with room temperature air.

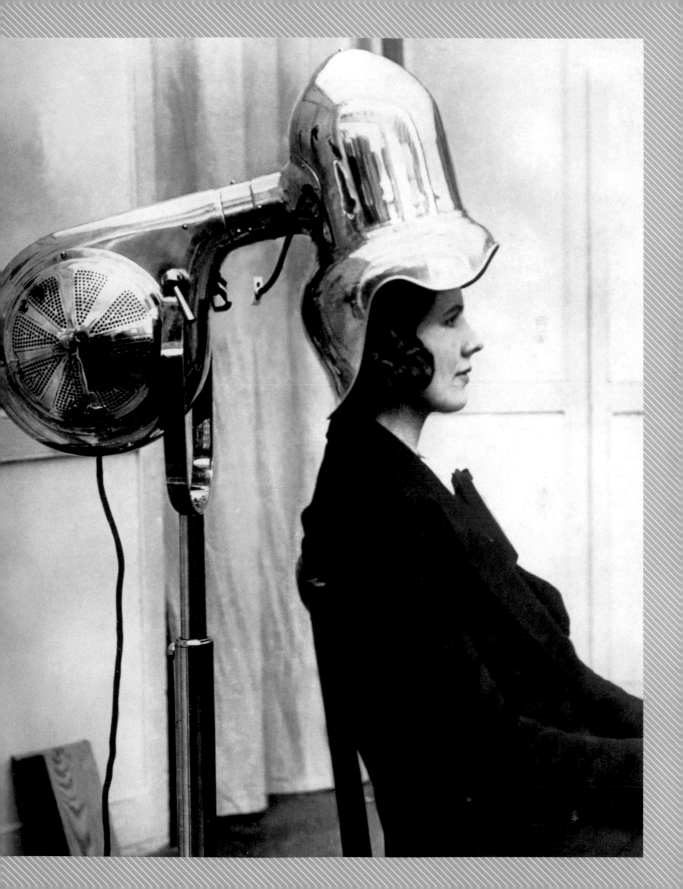

TAPE RECORDER

INVENTOR VARIOUS | 1930s

IT'S DONE WITH MAGNETS. That's the secret that early recording pioneers discovered as they developed ways for the human voice and other sounds to be copied and replayed. German scientists showed off the "Magnetophone" at the 1935 Berlin Radio Fair. A year later a London Philharmonic concert was presented on radio, even though it had been (gasp!) recorded earlier.

That was reel-to-reel tape, which became a recording-industry staple but proved awkward for home use. Cassettes and 8-track tapes became rivals in the mid-'60s with 8-tracks seeming to have the advantage, thanks to its ability to skip songs. But while the format had some sound advantages, it sometimes required the awkward rearrangement of songs from an album, occasionally interrupting a song midway through to switch tracks.

Cassettes eventually won, ultimately outselling albums—and giving rise to the mixtape, a boon to teen romance everywhere. The tape recorder also earned a spot in U.S. political history: Between 1971 and 1973, Richard Nixon used several Sony TC-800B machines to secretly record 3,700 hours of phone calls and meetings. Most controversial were the 18.5 minutes of tape allegedly erased as part of the Watergate coverup.

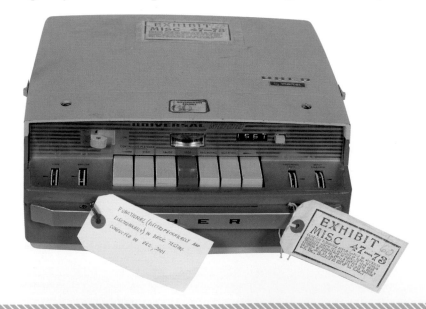

55

CB RADIO

INVENTOR FEDERAL COMMUNICATIONS COMMISSION (FCC) | **1945**

RARE IS THE TECHNOLOGY that emerges from a niche culture, becomes widely popular, and then retreats back to that same niche culture. That's just one of the unique things about the CB radio.

In 1945, a personal radio service regulated by the FCC hit the airwaves, permitting citizens a radio band for personal communication. Nobody paid much attention until 1958, when two-way communication became possible. The second booster rocket for the CB business came during the 1970s oil crisis, when big-rig drivers used the radio to alert each other of speed traps after the imposition of the national 55-mph limit.

The CB radio finally went mainstream with the 1975 chart-topping novelty song "Convoy" and the **1977 hit movie *Smokey and the Bandit*, starring Burt Reynolds** as a betting man trying to illegally transport Coors beer from Texas to Georgia.

The quills and fountain pens of yore were hollow, requiring non-clogging (read: slow drying and often messy) India ink to make the journey from writing utensil to page.

Laszlo Biro, a Hungarian newspaper editor, gets credit for the creation of the ballpoint pen, though it certainly helped having a chemist brother. Their breakthrough: using a pressurized tube rather than relying on gravity. After fleeing the Nazis, Biro and his sibling set up shop in Argentina, selling the pens under the name Birome. A big part of their success can be attributed to the Royal Air Force, which placed the first bulk order—over 30,000 pens to be used in flight.

Others got into the action, and, for a while, pens were luxury items. In 1945 America's first ballpoint pen, the Reynolds Rocket, cost $12.50—the equivalent of $150, or 546 Bic pens, today.

◀

Ballpoint technology kept the air out, which prevented ink from drying before it hit the paper. The ball also did double duty, rolling the ink out onto the page, like a tiny roll-on antiperspirant.

53 | CAR JACK

INVENTOR RICHARD DUDGEON | 1850

MOST GADGETS IN THIS BOOK are used—or were used—on a fairly regular basis. Then there are those we hope we *don't* have to use. So sits the car jack, quietly waiting near the spare tire in your car's trunk for its call of duty.

Next time you find yourself on the side of the road changing a flat, think of Richard Dudgeon, the guy who helped get the corner of your car off the ground. **He devised his first hydraulic jack in 1850, but initially used it in shipyards and railroad repair shops.**

Despite an increase in ease of use, today lots of people would rather someone else did the dirty work: In 2010 AAA came to the aid of over 3.7 million flat-tire sufferers.

Of course, none of this should be confused with carjacking, a crime first committed in 1912 but a term not commonly used until the 1990s.

52 | E-READER

INVENTOR VARIOUS | 1998

"I'M NOT A GADGET PERSON AT ALL, but I have fallen in love with this thing," said Oprah Winfrey, shortly after the Kindle's 2007 launch. And she's not alone.

The new gadget wasn't immediately embraced, but e-reading took off once readers realized they could carry hundreds of books on one device and easily purchase more.

But the Kindle wasn't the first e-reader on the market. NuvoMedia claims to be the first to get its version, Rocket eBook, in stores after cutting electronic rights deals with publishers in 1998. Initial models held around 4,000 pages—or about 10 novels—and had to be hooked up to computers to download books. But that soon changed, thanks to high-speed wireless.

The e-impact on the publishing industry directly compares to how iTunes impacted the music biz. The big difference? Publishers have been quick to turn the page, embracing the new revenue stream.

RELATED

1950s

LEAF BLOWER
Page 71

Although the lawnmower didn't catch on until the mid-20th century, patents show up as early as 1830. That's when Edwin Budding, a British machinist, crafted a machine "for the purpose of cropping or shearing the vegetable surface of lawns, grass plots and pleasure-grounds."

Budding's device had the recognizable elements of later push-mowers—a wheel that turned a cylinder with attached blades. But it didn't take off. Neither did the one Elwood McGuire invented in 1875 that used the wheels to turn the cutter.

That's because there were few actual lawns. The average person with land was more interested in growing fruits and vegetables than purely decorative grass.

With the end of World War II, though, came the boom in suburban living and its large manicured yards. Lawns were kept impeccably clipped, first by the descendants of Budding's and McGuire's creations and then by their noisy grandchildren, the power mower and the riding mower.

Edwin Budding got the idea for the cylinder mower from a cloth-trimming machine at a local mill. Mowers may not look like this anymore, but they essentially work the same way.

SO MUCH MORE CONVENIENT THAN THE LONG-PLAYING RECORD—AND ABLE TO WITHSTAND BOUNCE WITHOUT MISSING A BEAT—THE CASSETTE TAPE'S POPULARITY WAS FUELED BY THE AUTOMOBILE, THE BOOMBOX, AND THE WALKMAN.

The technology isn't so different from reel-to-reel tape. With a roll of 3.81-mm magnetic tape pulled between two reels and encased in plastic, the cassette puts tape into contact with a head in the player, which reads its signal.

Valdemar Poulson, a Danish telegraph worker, discovered the possibilities of magnetic recording, and early cassette players, the Magnetophon KL and Magnetophonband Type C, were shown at a Berlin exhibition in 1935. RCA was the first company to put reel-to-reel tape in cartridges. Others tried to get into the game, but competing formats proved too much of an obstacle. By 1965, though, the recording system was universal. Fun fact: Early cassettes had sticker prices comparable to some CDs today ($9.95—in 1950s dollars—for an hour of recorded time).

The cassette fell out of favor with the advent of the more durable compact disc. Not only could CDs be made more cheaply, the CD player manufacturing itself was more cost-efficient. Another bonus: You didn't have to guess when you wanted to skip your least favorite song on an album.

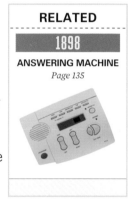

RELATED

1898

ANSWERING MACHINE
Page 135

49 CIRCULAR SAW

INVENTOR EDMOND MICHEL | 1923

t's easy to imagine someone observing the power of spinning blades in a sawmill and saying, "Hey, why don't I have one of those in my garage?"

Who had that thought first—and acted on it—remains unknown. Claims for the creation of the circular saw come from Germany, England, and Holland. But we give credit to Edmond Michel, a Frenchman living in New Orleans.

His early versions married a machete (the sugarcane chopping weapon of choice) with a malted-milk-mixer blade and a 2-inch circular blade. Add a shaft and a gearbox and, voilà!: A slow and impractical device that nonetheless pointed the way to today's circular saw.

The Michel Electric Hand Saw tested on both coasts, including work on New Jersey boardwalks. After Michel opted out of the company, Skilsaw became its moniker—and the name by which the saw, even when produced by others, became known.

The device didn't just make cutting easier. It also paved the way for the Shop-Vac.

GAME BOY

INVENTOR NINTENDO | 1989

Among the most successful gaming systems ever—118 million units and a half-billion games sold—the Game Boy has been making road trips more peaceful for parents since 1989.

Given the portability of audio and video products at the time, portability of video games seemed inevitable. Despite lots of single-game-dedicated devices, the big money came from finding a system that could play many games—and require loyal players to buy updated machines when they stopped being the latest and greatest.

Game Boy offered a 2½-inch liquid crystal display screen with a few essential controls—a black directional cross, two

round buttons, and two slash buttons. Home video game players knew not to expect the same quality graphics they were used to on their TVs or computer screens. The first Game Boy games were one-color with images often difficult to see, but the device did have Tetris bundled into the package.

Players could also link to other Game Boys for one-on-one competition. With that feature, at least parents could pretend that human interaction occurred.

RELATED

1816	1939
KALEIDOSCOPE *Page 18*	**VIEW-MASTER** *Page 8*

Frustrated on a European vacation when a pocketknife couldn't turn a stripped radiator handle, engineer Tim Leatherman invented the product that now multitasks around the world. What made the device worthy of patent?

The dual-action pliers combined with an integrated locking clamp.

After being rejected by manufacturers, Leatherman eventually formed Leatherman Tool Group and introduced the Pocket Survival Tool in 1983. First sold through mail-order catalogs, the Pocket Survival Tool had 14 gadgets, including pliers, four screwdrivers, a can opener, an awl, and a blade.

Retired in 2004, the Pocket Survival Tool led to many variations, including Flair (containing a corkscrew and a cocktail fork), Juice (which colorized the casing), Core (with hollow-ground screwdrivers), Skeletool (not really innovative, but with a cool name), and the military-friendly Mut (with a sheath and scope-adjustment wrench).

It certainly helped, marketing-wise, when a Leatherman appeared on a 1999 episode of the sci-fi hit series *The X-Files*. After all, you never know when you're going to need to extract alien flesh.

> **RELATED**
>
> **1891**
>
> **SWISS ARMY KNIFE**
> *Page 74*

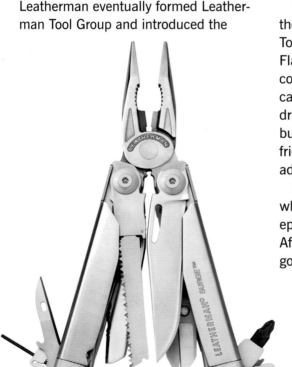

BE PREPARED
Tim Leatherman envisioned his multi-tool as a "Boy Scout knife with pliers." Most Leatherman tools are built around full-size pliers and offer up to 21 additional tools.

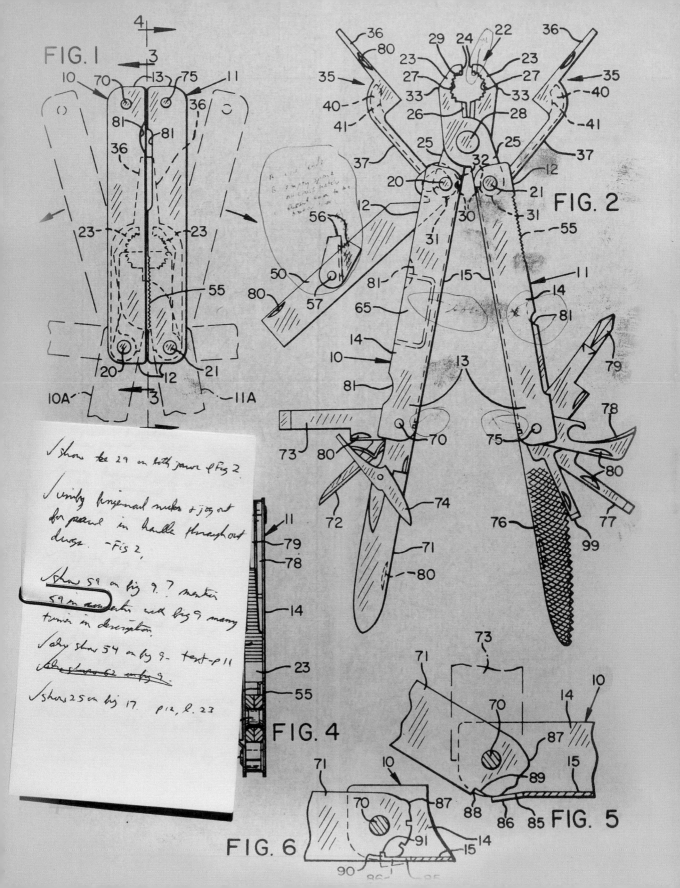

THE FIRE EXTINGUISHER wasn't designed to put out fires, it was intended to limit the spread of flames until the professionals arrived.

The first model, dating back to 1723, literally fought fire with fire. Imagine a cask that held water with a pewter gunpowder chamber connected to fuses. In case of fire, the occupant lit the fuses to ignite the gunpowder, which exploded, scattering the water.

Thankfully, the technology has improved since then, starting with British Captian George William Manby's 1818 invention that incorporated compressed air. The soda-acid extinguisher was patented in France in 1866, followed by the cartridge-operated version in 1881. Chemical foam joined the party in 1904, thanks to Russian chemist Aleksandr Loran.

Today most fire extinguishers use dry chemical powder, although others use pressurized water or carbon dioxide. (Halon, a once-popular retardant, isn't found much now, as evidence shows it depletes the ozone layer.)

All fires are not created equal, so different chemicals are needed to douse them. It's not a good idea to use a carbon dioxide extinguisher to handle a fire of Class A materials (wood, cloth, or paper). Similarly, a dry chemical extinguisher isn't going to help a Class K (cooking oils and animal fats) inferno.

Today, U.S. homeowners **douse 13 million fires annually** with extinguishers that spray foam, monoammonium phosphate, or sodium bicarbonate.

RECORDED MUSIC was originally consumed in the home. Then its reach expanded to the automobile and then, through the boom boxes of the 1970s, out into the world. But music didn't really become personal until the invention of the Walkman.

Sony launched the Walkman, a name credited to engineer Nobutoshi Kihara, in 1979. The tape player could be hooked to your person and worn

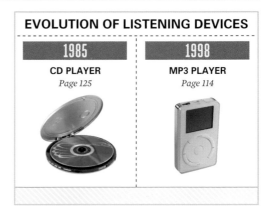

EVOLUTION OF LISTENING DEVICES

1985	1998
CD PLAYER *Page 125*	**MP3 PLAYER** *Page 114*

anywhere, preferably with headphones. Finally, it was possible to hear the music of your choice on the exercise bike, the city bus, or the production line. **By 1995 the Walkman had sold more than 150 million units.**

But the Walkman never would've had legs without German-born Brazilian inventor Andreas Pavel and the Stereobelt he patented in 1972. After a long legal battle, Pavel successfully setted with Sony in 2004.

44 FLOPPY DISK

INVENTOR IBM | 1971

EVER WONDER WHAT THAT ICON THAT ACCOMPANIES "SAVE" ON YOUR COMPUTER'S DROP-DOWN MENU IS?

You, my friend, are young. Or at least, you weren't of computing age in the 1970s and '80s, when the floppy disk was the computer storage method of choice for material that would clutter your computer's hard drive or need to be moved from one computer to another.

Early adopters loaded operating systems and programs via such disks, **which IBM introduced in 1971.** Floppies kept evolving— from 8 pliable inches to 3.5 rigid ones—but the rise of CD-ROMS insured their extinction.

It's a safe bet that any computer user from that time still has some important files secured on floppies. Too bad he'll never be able to read them.

EVOLUTION OF DIGITAL STORAGE

1988	1964	1999
CD-ROM	**PORTABLE HARD DRIVE**	**FLASH DRIVE**
Page 94	*Page 46*	*Page 64*

POLAROID CAMERA

INVENTOR EDWIN LAND | 1946

While watching her father Edwin take photos in 1943, three-year-old Jennifer Land complained, "Why can't I see the pictures now?" Why indeed. In 1948 Polaroid—the company her father headed—began selling instant film and the first commercial instant camera, the Land Camera.

More than 10 years and $250 million were spent developing the iconic SX-70 model, which debuted in 1972. The expenditure nearly sank the firm, but by 1974 the camera was a hit: It spit out 1 billion prints that year.

The camera itself was clunky, but that's the price you pay to have photos in your hands instantly. As Dr. Land eloquently said of the process, it allowed the camera "…to generate a new intimacy between the human being and the world around him…and then, looking through the camera, go quickly from focusing to touching the button…and capture that feeling you had."

EVOLUTION OF THE CAMERA

1900	1971
BROWNIE CAMERA	DIGITAL CAMERA
Page 138	*Page 120*

42 MP3 PLAYER

INVENTOR VARIOUS | 1998

Before the MP3 we stored our music on cassettes or CDs. Then along came a way to carry thousands, rather than dozens, of songs conveniently, and those cassettes began to look like ancient technology.

It wasn't embraced by all. MP3 marked either the end of civilization (record companies) or the dawn of a new world (everyone else). When the first commercially viable MP3 player, the Diamond Rio, hit the shelves in 1998, the Recording Industry Association of America sued—providing massive publicity and a boost to digital technology. And the 1999 launch of Napster, essentially a centralized file server for music sharing, led to years of litigation and resistance from the music industry.

The MP3 era didn't truly start, though, until the 2003 launch of Apple's iTunes store, two years after the first iPod was sold. iTunes changed the music industry, allowing the purchase of single tracks—a form once thought dead. The outlet has since sold more than 10 billion songs.

Music would never be the same. The fallout hit the radio business (which lost its hold on in-car music), music retailers (selling a product that could be purchased instantly from a computer), and the singers and songwriters themselves (who began focusing on hit singles, instead of albums).

All of which explains why the average American's music spending dropped to a mere $26 in 2009, while personal music libraries grew exponentially.

EVOLUTION OF LISTENING DEVICES

1979	1985
SONY WALKMAN	**CD PLAYER**
Page 111	*Page 125*

41 | POCKET CALCULATOR

INVENTOR TEXAS INSTRUMENTS | 1965

MATH NERDS COULDN'T SLIP the first all-transistor calculator (invented in 1957) into their shirt pockets for a very good reason: The three-unit IBM 608 weighed 2,400 pounds. Not to mention that at $83,210, it was far too pricey for your average geek.

Let's skip past the multi-pound devices—as well as those with outputs via roll paper—to the early 1970s when HP introduced the first calculator that went beyond the four basic functions of addition, subtraction, multiplication, and division. By 1976 four-function pocket calculators weighed a few ounces and cost a few dollars—a price so low that many companies in the business looked to create scientific and programmable calculators they could sell for more.

In 1985 Casio introduced the first graphing calculators. They quickly became a requirement in advanced high school and college math classes—and a new way for students to play bootleg versions of *Tetris*.

40 | WI-FI ROUTER

INVENTOR VARIOUS | 1999

THE FIRST STANDARDS FOR WI-FI were developed in 1999—when the few people who heard the term probably thought they were mishearing "hi-fi." The following year, the first Wi-Fi products hit the market, revolutionizing the way we access data around the world.

Wi-Fi—making use of the airwaves—allows multiple devices to be connected simultaneously. Gre Ellis, one of the Wi-Fi pioneers, says it's more of a series of "protocols that are used so that everyone is sharing the airwaves properly."

Wi-Fi has made its way into more than 9,000 devices, from phones to TVs, since its introduction in 2000. According to a Wi-Fi Alliance poll, 75 percent of young Americans say they'd give up coffee before Wi-Fi. Receivers such as the Novatel MiFi hotspot, which takes long-range 3G and 4G cellular signals and turns them into short-range Wi-Fi networks, may eventually allow high-speed Internet access to reach areas where broadband infrastructure cannot.

39 ELECTRIC DRILL

INVENTOR JAMES ARNOT | 1889

Drills of one kind or another have been part of the toolbox for almost as long as there have been toolboxes. And with each version, the tool has become easier to use. As a result, Black & Decker introduced the workshop's common power tool in 1916, with a grip loosely based on the handle of a Colt .45.

Australian inventor James Arnot patented the first electric drill in 1889 and German inventor Wilhelm Fein created a hand-held version in 1895, but heavy shells made these cumbersome. Switching out steel for aluminum made them far more practical. And swapping the nickel-cadmium batteries for lithium-ion batteries allowed for longer use between charges.

38 DIGITAL HDTV
INVENTOR NIPPON HOSO KYOKAI (NHK) | 1979

IN 1987 WHEN JAPANESE ENGINEERS showed off their MUSE hi-def system, policymakers in Washington, D.C., responded by pushing for new broadcast standards. After many delays HDTVs finally arrived in stores in 1998.

Before that year TV display screens had a common width-to-height ratio of 4 to 3. The picture itself consisted of 525 horizontal line scans with dots of color—pixels—populating each scan line, the number of which varied from set to set. Digital television not only brought a clearer picture, it upped the quality of the sound; reframed TVs so that wide-screen movies could be watched without letterboxing; and allowed for more creative ways to rewind, record, and fast-forward.

The Consumer Electronics Association didn't even start tracking HD sales until 2003, when the average set cost over $1,600. Today, they're in more than half of U.S. households.

37 WRISTWATCH
INVENTOR LOUIS CARTIER | 1904

IN 1901, WHILE CELEBRATING IN PARIS after winning a prize for circling the Eiffel Tower in a dirigible, Brazilian aviator Alberto Santos-Dumont asked his friend Louis Cartier to design a watch that would permit him to time his aerial maneuvers and still keep his hands on the controls.

Santos-Dumont was wearing this new device on October 23, 1906, when he made the first successful airplane flight in Europe—and the first flight anywhere in a craft with nondetachable landing gear.

In 1911 wristwatches were made available to the public, and soon they were everywhere. Self-winding watches hit the market in the 1920s, followed by electric-powered watches in the 1950s. The quartz watch premiered in Tokyo in 1969, where it sold for about the same price as a family car.

36 8-MM CAMERA

INVENTOR EASTMAN KODAK | 1930s

Never practical for large professional film productions, the 8-mm camera (and its sequel, Super-8) provided the most cost-effective way to record and preserve moving images of family members—even if there was no audio. Film buffs embraced the format, owning home versions of their favorite silent films in 4- or 12-minute reels.

Eastman Kodak first developed the 8-mm moviemaking format in the 1930s. Its Cine Kodak Eight was actually 16-mm film with sprocket holes on both sides. After running through the camera one way, it was reloaded to go through again. The processing lab would do the cutting and splicing to make one continuous series of frames.

By the 1980s, though, most talk of 8-mm and Super-8 had to do with methods for transferring those images to the then-dominant VHS format. Even when J.J. Abrams directed the science fiction film *Super 8*, he could only bring the format back as an object of nostalgia.

Perhaps the most famous home movie in history was caught in the format: 486 frames on Kodachrome II safety film that capture the assassination of President John F. Kennedy. (The filming almost didn't happen—Abraham Zapruder had left his Bell & Howell 414PD Zoomatic camera at home that day, but at his assistant's urging, drove 14 miles round-trip to retrieve it.)

RELATED

1983

CAMCORDER
Page 126

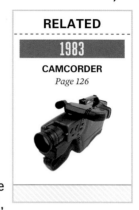

A MICROPHONE DOES TWO PRIMARY THINGS.

It turns sound into electronic impulses that can be transmitted or recorded. And it gives weak sounds a boost—(micro+phone)—which is how it got its name.

From the early telephone transmitters to the nearly invisible miniature mics today, sound-capturing tools are as essential to mass culture as those that capture images. Without the microphone, you might still have movies and TV, but you wouldn't be able to hear them. And as with any evolving gadget, individual designs can take you back to a fairly specific time and use. Put an RCA 44 Style Ribbon Microphone or a Shure Unidyne Model 55 in a film, and you instantly know you're in the 1930s or '40s.

Other long-term side effects of microphone technology range from bringing down a U.S. president (goodbye, Richard Nixon) to bridging of the cultural gap between East and West (hello, karaoke).

34 | DIGITAL CAMERA

INVENTOR STEVEN SASSON | 1975

THE FIRST DIGITAL CAMERA—AN 8-POUND, 0.01 MEGAPIXEL DEVICE—WAS A COMBINATION OF NOT ONE, BUT TWO ANTIQUATED TECHNOLOGIES.

"It had a lens that we took from a used parts bin from the Super 8 movie camera production line downstairs from our little lab on the second floor in Bldg 4," recalled Kodak engineer Steve Sasson. "On the side of our portable contraption, we shoehorned in a portable digital cassette instrumentation recorder." (Not to mention the 16 NiCad batteries it required.) Black and white images were recorded to a cassette and then displayed on a TV set.

Many innovations later, the digital camera did for the standard film camera what video did to the Super-8 camera. And it made it popular to keep shooting until you got something not half bad.

The earliest buyers of digital camera were rarely "photographers" in the arty sense of the word.

EVOLUTION OF THE CAMERA

1900	1946
BROWNIE CAMERA	**POLAROID CAMERA**
Page 138	*Page 113*

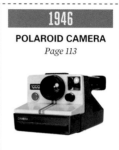

They were generally business-users, who found that the huge savings in film and printing—plus the ability to transmit images from remote locations—easily offset the high cost of the cameras themselves. Magazine art directors found they didn't always need to pay top shooters for simple jobs. Realtors, insurance adjusters (as well as accident victims), and auction-house owners could quickly inventory lots of reference photos. You could even take your own passport photo, deleting those that made you look like you should be on a watch list.

But there were victims: Because of the popularity of this gadget, introduced in 1990, Kodak retired its Kodachrome film format in 2009, after 74 years of service.

33 | MICROWAVE OVEN

INVENTOR PERCY SPENCER | 1945

IN 1945 RAYTHEON'S PERCY SPENCER stood in front of a magnetron, the power tube of radar, and felt a candy bar start to melt in his pocket. Intrigued, he then placed popcorn kernels in front of the magnetron, and the kernels exploded all over the lab. (Yes, the first item cooked by a microwave on purpose was microwave popcorn.)

Part of the galley of the nuclear-powered passenger/cargo ship the NS *Savanna*, Raytheon's first "radar range"—which cooked with high-frequency radio waves—was water cooled, stood nearly 6 feet tall, and needed significantly more power than the one in your kitchen.

The development coincided with the women's rights movement. As two-worker families became the norm, so did expedited meal preparation. By 1986 microwaves were in nearly a quarter of U.S. homes.

Was Amana overstating the case when it called the microwave "the greatest cooking discovery since fire"? Try to imagine your kitchen without one and then decide.

▶ Douglas Engelbart wasn't the first to conceptualize a rolling device to manipulate a computer cursor. In 1952 researchers used a bowling ball to create the first trackball for the Royal Canadian Navy.

But computer history paints Englebart of the Stanford Research Institute as the father of the device. He was actually demonstrating videoconferencing and hypertext links at a San Francisco computer conference in 1968, but it was his one-button rolling wood block that caught the audience's imagination.

The mouse didn't really become practical, though, until Apple paired a one-button, steel-ball version, about the size of a deck of cards, with its Lisa computer in the early 1980s. Steel quickly gave way to plastic. Balls gave way to optical and then integrated mice.

And why was it named after a rodent? Doug Engelbart says, "I don't know why we call it a mouse. It started that way, and we never changed it."

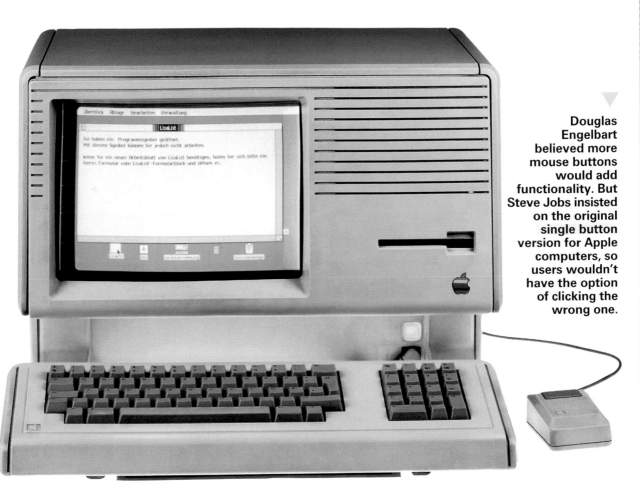

Douglas Engelbart believed more mouse buttons would add functionality. But Steve Jobs insisted on the original single button version for Apple computers, so users wouldn't have the option of clicking the wrong one.

EVOLUTION OF THE PERSONAL COMPUTER

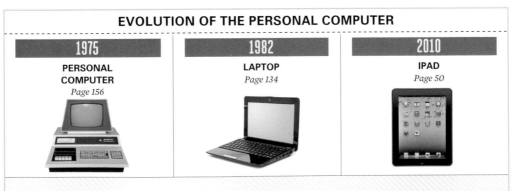

1975	1982	2010
PERSONAL COMPUTER *Page 156*	**LAPTOP** *Page 134*	**IPAD** *Page 50*

GOODBYE TRADITIONAL LIGHTBULBS, YOU OLD-SCHOOL ENERGY HOGS. YOU JUST CAN'T BEAT THE EFFICIENCY OF LEDS (LIGHT-EMITTING DIODES).

With an average life span six times longer than compact fluorescent lamps and about 50 times that of incandescents, they use less than a third of the energy, with significant cutbacks in CO_2 emissions to boot.

Commercial light-emitting diodes were first used in Hewlett-Packard calculators in the 1960s, and in the '70s Fairchild Optoelectronics introduced a process that reduced the cost to pennies apiece. Early versions of LEDs found their way into pioneering handheld calculators before becoming ubiquitous in radios, TVs, telephones, and watches.

These days you could probably find your way around your house in the middle of the night by LED lights alone. And with the government mandated phase-out of old-fashioned incandescent lightbulbs by the end of 2013, the push for scientists to develop light white enough for general use became more of a shove. The development of quantum dots—tiny semiconductor crystals that produce a whiter light—earned developers a Popular Mechanics Breakthrough Award in 2006.

◄

Since the 1960s, LED efficiency and light output has doubled about every 36 months. With a lifespan of 50,000 to 100,000 hours, many LEDs from the '70s and '80s are still going strong today.

30 CD PLAYER

INVENTOR JAMES RUSSELL | 1985

THE TRANSITION FROM CASSETTES TO CDs WASN'T EASY. Retailers had to be convinced that double stocking would pay off. Luckily it did, to the tune of 20 billion CDs sold by 2007. This also marked the era of a new age: the first step between analog and digital media.

James Russell gets credit for this gadget, landing 26 patents for CD and CD-ROM technology after realizing that binary notation (zeros and ones) could be optically translated to light and dark, thus becoming readable to a laser or computer.

The benefits were enormously marketable. First, there was no mechanical contact between the disc and the reader, so no chance of damage to the disc. And even if minor glitches occurred through handling, error correction bits eliminated that scratching sound all too familiar to LP listeners. The small size (with greater music-carrying capacity than LPs) combined with non-contact performance meant a smaller

EVOLUTION OF LISTENING DEVICES

1979	1998
SONY WALKMAN *Page 111*	**MP3 PLAYER** *Page 114*

player. That, in turn, broke open the technology for car stereo use and listening on the go. Thus the Walkman gave way to the Discman.

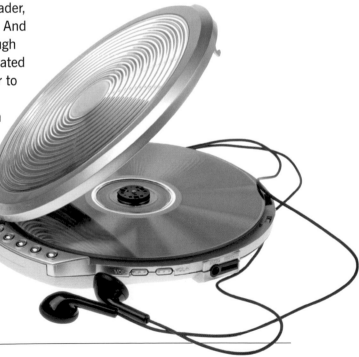

▶ **Sony's first portable CD player, released in 1984, was called a Discman. In the early '90s, the company re-named the device the CD Walkman.**

㉙ CAMCORDER

INVENTOR JVC/SONY | 1983

EXPENSIVE AND FLAMMABLE. Those were two of the main reasons why 35-mm film—the standard for theatrical motion pictures—didn't cut it as a home movie format. The Cine Kodak camera's 16-mm proved a bit safer, but had to be hand-cranked during filming. Then came 8-mm and its sequel, Super 8, which halved the size of the film and cut the amount of celluloid necessary by a quarter.

RELATED

1930s

8-MM CAMERA
Page 118

In 1983 both JVC and Sony released camcorders, combining camera and recorder. The problem? Different formats. After a fierce and bloody battle, the favorite—Betamax—lost to the larger VHS. And the world

suddenly became a more recorded place. Family vacations meant carting around a lightweight camcorder (using the '80s definition of lightweight, of course). Wedding ceremonies required play-by-play permanent records. And at the height of the show's success, producers of *America's Funniest Home Videos* sorted through 2,000 tapes a day to find optimal moments of non-fatal collisions, awkward animals, and accidental kicks to the groin.

28 | ELECTRIC GUITAR
INVENTOR **GEORGE BEAUCHAMP** | 1931

It can scream with ear-shattering intensity or whisper gentle melodies, sear with a torrent of winding scales or dominate a space with massive, organ-like chords. It has shaped the sound of popular music since the 1950s—from jazz and country to rock and folk. And there are a few people today who haven't at one time or another tried—or at least yearned—to play one.

The names of the pioneers in the development of the electric guitar—Rickenbacker, Fender, and Les Paul—are revered by those who play and are known even by casual fans.

Rickenbacker (with George Beauchamp) first used electricity to amplify the string's vibrations. Les Paul mounted strings and pickups on a block of pine, and gave the guitar world overdubbing, reverb, echo, and a lot more. And Leo Fender designed the first mass-produced solid-body electric guitar, the Telecaster, in 1951—when Keith Richards was just a lad of 8.

Without it, Jimi Hendrix's version of "The Star-Spangled Banner" would sound more like everyone else's; Bob Dylan wouldn't have upset the crowd at the 1965 Newport Folk Festival; and Spinal Tap wouldn't need a speaker that went up to 11.

STILL IN TUNE

While amplification equipment has advanced, today's musicians—the ones that actually play instruments, that is—rely on the same electric guitars first used in the 1950s.

(27) | BLACKBERRY

INVENTOR MIHALIS LAZARIDIS | 1999

Now in fourth place behind the iPhone, Android, and Windows Phone, BlackBerry still attracts many users worldwide with BlackBerry Messenger, which sends encrypted instant messages, voice memos, videos, and pictures to other Black-Berry users for free.

"IT WILL SOON BE POSSIBLE to transmit wireless messages so simply that an individual can carry and operate his own apparatus," said Nikola Tesla in 1909. He called it. Flash-forward 90 years—not a long time, really—and the BlackBerry arrived.

Research in Motion (RIM)—headquartered in Waterloo, Canada—didn't get as much attention as Microsoft or Apple—but for a while, its prime product proved as much a corporate status symbol as an increasingly essential device. Initially the moniker of a two-way pager, BlackBerry soon found itself repositioned as a pioneering smartphone. Led by Mihalis "Mike" Lazaridis, RIM initially played up BlackBerry's strength in email, pretty much owning the market during crucial years. Post-9/11 the U.S. government become the biggest BlackBerry customer, with half a million units being thumb-controlled. And the term CrackBerry became part of the lexicon, referring to those who compulsively checked their mobile devices.

Today we just call them people.

Ask for a Kleenex and you might be handed any facial tissue. Ask for Rollerblades and you won't necessarily get inline skates sporting that brand name.

And just as with those two iconic brands, when you ask your mechanic's assistant to hand you a Crescent wrench, there's no guarantee you'll get one from the Crescent Tool Company slapped into your palm.

Crescent didn't invent the adjustable wrench—one that allows the angle and the width of the jaw to be changed and locked. It just popularized it to a point that it became ubiquitous in toolboxes. In some European countries, it's still known as an English key, giving credit to the home country of Richard Clyburn, inventor of an early version.

Designed primarily as a tool for motorists who needed to make frequent adjustments to the brakes and clutches of early autos, the original Crescent wrench replaced the box of fixed-size wrenches needed to deal with varied nut and bolt sizes.

FLYING COLORS

Of course, in turning an everyday tool into a brand icon, it helps to have a high-profile endorser. On his 1927 trans-Atlantic flight, Charles Lindbergh said he carried "gasoline, sandwiches, a bottle of water, a Crescent wrench, and pliers." You can't buy publicity like that.

IT'S A NOUN. IT'S AN ADJECTIVE. It's a marketing move. Hi-Fi entered the vernacular in the 1950s, when the novelty of hearing recorded or broadcast sound had long since worn off. It wasn't enough just to hear the sound—the emphasis was on the quality.

Suddenly gadget buffs were masquerading as music purists, pridefully touting the audio benefits of high-fidelity recordings and equipment over old-school 78-rpm records and AM radio. FM had a dramatically improved sound quality. Thirty-three and one-third rpm recordings had less surface noise. Jazz and classical music had an intricacy that those barbaric older forms insulted.

But, of course, you needed better equipment.

To achieve maximum sound quality, audiophiles and, later, mainstream equipment buyers, started assembling their sound systems in components, with separate tuners, amplifiers, speakers, and turntables. The end result focused on sound that seemed second to none—until stereo sound became the newest and greatest.

○ MUSIC MACHINES

THIS GRAPH TRACKS THE RISE AND FALL OF FORMATS BY ANALYZING
HOW MUCH THE AVERAGE AMERICAN SPENT ON MUSIC PER YEAR.

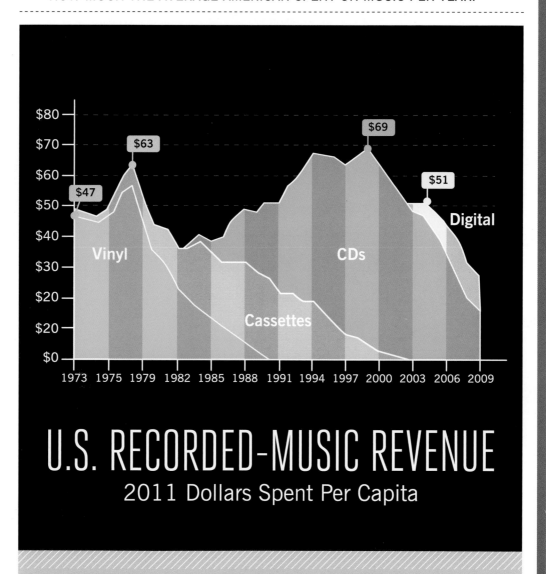

U.S. RECORDED-MUSIC REVENUE
2011 Dollars Spent Per Capita

From the 1920s until the '80s, consumers spent the most per capita on vinyl records for phonographs
(No. 8) and hi-fi (No. 25). Cassette tapes (No. 50) spooled songs on the Walkman (No. 45) and
boombox (No. 70). Spending on compact discs, spun on CD players (No. 30), overtook tapes in the
early '90s. Digital piracy and single-track downloads to the iPod (No. 42) explain why per capita
spending is down—and the music industry in a tailspin.

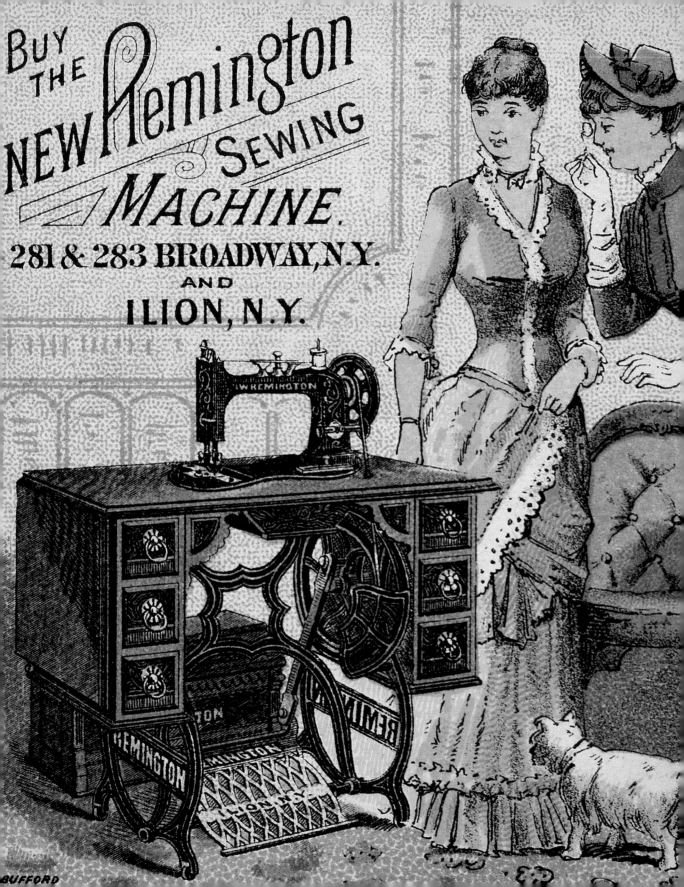

SEWING MACHINE

INVENTOR ISAAC MERRITT SINGER | 1933

THE FIRST MECHANICAL STITCHER CUT THE TIME IT TOOK TO SEW A SHIRT FROM 14 HOURS DOWN TO JUST 1 AND

turned a home-based activity into an industry. (Early machines, made of cast iron, used a foot pump—a design element that lives on to this day as the food pedal.)

Eighty years later, Singer Manufacturing Company unveiled the 11-pound Featherweight, a portable version of the sewing machine, at Chicago's 1933 Century of Progress fair. Complete with an extension table and easy thumbscrew release for lubrication, the device was the latest in a long series of labor-saving developments. Company founder Isaac Merrit Singer's work was built on processes developed by, among others,

Thomas Saint in the 1790s and Elias Howe in the 1840s.

Singer patented his improvements—including having the shuttle go in a straight line instead of a circle—and pooled patents with other innovators, including Howe. Mass production began. So did mass advertising, as Singer needled homemakers into believing that his device was an essential tool for every household.

But the influence of the sewing machine went well beyond the home. Factories, including exploitative sweatshops, were built around the device. On the other hand, without it, we wouldn't have *Project Runway*.

LAPTOP
INVENTOR **VARIOUS** | 1982

IN THE AGE OF THE MACBOOK AIR, a computer weighing more than a few pounds feels as bulky as a desktop. So take the game-changing early laptop, the 11-pound GRiD Compass 1101, with a grain of historical salt. This clamshell computer, invented by William Moggridge and others, went on sale in 1982. Like other early '80s models, such as the 29-pound Kaypro II, it wasn't something you toted around lightly.

But it proved a boon to college students, work-at-homers, and *Tetris* addicts. By 2011, 52 percent of U.S. adults owned a laptop—among them, seven in 10 adults ages 18 to 34.

Plummeting prices certainly helped. In 2000 a Gateway laptop would have set you back $1,300, despite that fact that it included a year of free—yes, FREE—AOL. In a decade the price had halved

and the processing power jumped from 550 MHz to 2.13 GHz. No matter when you became a laptop user, the first one no doubt seems like a suitcase compared to the one you have now. Unless you're already abandoned yours for a tablet.

EVOLUTION OF THE PERSONAL COMPUTER

1968	1975	2010
COMPUTER MOUSE *Page 122*	**PERSONAL COMPUTER** *Page 156*	**IPAD** *Page 50*

22 VCR

INVENTOR VARIOUS | 1975

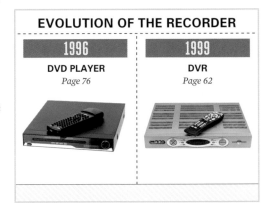

"NOW YOU DON'T HAVE TO MISS *KOJAK* because you're watching *Columbo*," boasted Sony's 1976 ad for the first Betamax Video Cassette Recorder.

The original Betamax VCR would record a single hour of moving images on magnetic tape, accompanied by monaural sound with a good deal more hiss than the average phonograph record. Special features were limited: It only could be programmed in advance to record a single show—and all for a $3,000 price tag.

TV studio execs rushed to the courts to protect copyrights. In 1984, the Supreme Court okayed home taping, settling the issue.

EVOLUTION OF THE RECORDER

1996	1999
DVD PLAYER	**DVR**
Page 76	*Page 62*

21 ANSWERING MACHINE

INVENTOR VALDEMAR POULSEN | 1898

BACK IN THE OLDEN DAYS, when a phone rang, someone had to be there to answer it. Otherwise, that caller—whoever he or she may be—just had to ring again. People like doctors, lawyers, and very desperate actors used answering services staffed by proxies who would pick up calls from a remote office and relay messages. (See the 1956 Judy Holliday musical "Bells Are Ringing".)

Answering services could hear their demise when the Ansafone, the first commercially successful answering machine, hit the market in 1960. The PhoneMate Model 400 later became the first widely-used answering machine in 1971. Its tapes held up to 20 messages, enabling a new form of selective communication.

SEE ALSO

1898

CASSETTE TAPE
Page 105

▶ Despite how much time we spend looking for it, the TV remote control is a device we can't imagine living without.

The first remote, the aptly-named Zenith Lazy Bones, was invented in 1950 and tethered to the tube by a wire. Five years later, the first wireless version, the gun-shaped Flash-matic, operated the TV with a beam of light. "You can also shut off long, annoying commercials while the picture remains on the screen," (long, annoying) ads promised. With rough early technology, you could also make your TV go haywire.

Next came ultrasound, developed by inventor Robert Adler, who later patented his own wireless remote. Clicks from the remote were heard by the TV, which responded accordingly. Unfortunately, it also responded to other sounds, so a set of keys, a tinkling cat collar, or an ultrasonic burglar alarm caused unwanted channel surfing.

The technology continued to evolve. Emerson tried a model with a built-in speaker. Motorola's Transituner could only change stations. And RCA began pushing its Wireless Wizard circa 1960. Skipping stations wasn't possible until the '70s when Magnavox added a keypad to its remote. Infrared technology had its days with GE's remote that used a trio of pulsing LEDs to change the channel.

Even Apple cofounder Steve Wozniak had a piece of the action when his company CL 9 created the universal remote—originally priced at $200.

What's next? The smartphone remote control. Now that's universal.

RELATED

1930s

TELEVISION
Page 160

19 BROWNIE CAMERA

INVENTOR EASTMAN KODAK | **1900**

"EVERY TECHNOLOGY WE USE TO COMMUNICATE WITH PICTURES CAN TRACE ITS ANCESTRY BACK TO THAT FIRST BLACK BOX,"

EVOLUTION OF THE CAMERA

1946	1971
POLAROID CAMERA	**DIGITAL CAMERA**
Page 113	*Page 120*

Kodak noted on its website in celebration of the Brownie's 100th birthday.

That's no exaggeration. Inexpensive and easy to operate, the Brownie, one of the first box cameras, hit stores in 1900. About 100,000 Brownies were sold in its first year on the market. The only downside: It allowed for just six exposures without reloading.

The Brownie put photography into the hands of the masses and inspired no less than legendary photographer Ansel Adams, who at age 14 received one from his parents during a 1916 trip to Yosemite. While setting up his first photo, Adams tumbled off a tree stump and inadvertently pressed the shutter. He called the accidental image "one of my favorites from this, my first year of photography."

The compact camera was named after The Brownies, popular cartoon characters created by Palmer Cox, that would later be featured in ads targeting kids. The name was also a nod to camera designer **Frank Brownell.**

VACUUM CLEANER

INVENTOR JAMES SPANGLER | 1908

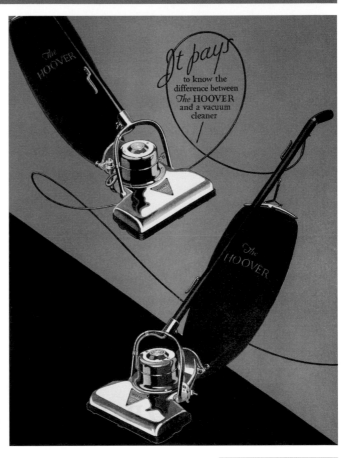

Basically a motor-driven fan with a nozzle attached, the vacuum cleaner uses atmospheric pressure to force air—and the dirt in it—first into a nozzle and then into a bag or storage container.

Like early models of most popular gadgets, the first vacuum cleaners were well beyond the financial reach of most people. The 1908 Hoover Model O went for $60, which doesn't seem like much until you consider that that's over $1,400 in 2013 dollars. By the mid-1950s prices dropped enough to make upright vacuums a middle-class staple—thanks in part to the pressure applied by door-to-door vacuum-cleaner salesmen. The question soon shifted from "Do you have a vacuum cleaner?" to "Bagged or bagless?"

Innovation continued with James Dyson inventing the cyclonic vacuum in the 1980s, and robotic vacuum cleaners jumping from science fiction novels to the living room in the early 2000s. Which, if you'll pardon the expression, doesn't suck.

RELATED

2002

ROOMBA
Page 55

17 HANDHELD GPS

INVENTOR UNITED STATES DEPARTMENT OF DEFENSE | 1973

THE TRUTH IS, BEFORE GPS, WE USUALLY GOT TO WHERE WE WERE GOING. BUT THE TOOL WE USED TO GET THERE WAS A PAIN TO FOLD.

GPS made getting around easier and covered more ground than a simple map ever could. In operation since the mid-'80s, it was originally used in U.S. military navigation to tell moving aircraft, warships, and missiles exactly where they were in respect to targets. The Reagan administration opened GPS up for civilian use in 1983 after the Soviets downed a Korean airliner in a no-fly zone. And because of the fear that it would be used for destructive purposes, data errors were intentionally included.

In a nutshell, GPS is an exercise in triangulation. Each satellite keeps track of two things—the location and time—and sends signals to Earth containing that information. GPS calculates the difference between signals from multiple satellites to pinpoint the location.

Magellan sold the first handheld unit in 1989. Less than a decade later, GPS units were available for under $100 and, like just about everything else, found their way onto smartphones. Since then "Recalculating" has become an in-car refrain, replacing the age-old question "Are we there yet?"

16 TRANSISTOR RADIO

INVENTOR TEXAS INSTRUMENTS | 1954

Until the 1950s the radio was anchored in the living room. That changed when vacuum tubes gave way to transistors.

The Regency TR-1, a collaboration between transistor maker Texas Instruments and radio maker Regency, was designed to bring attention to transistor technology. Tshushin Kogyo, the company that eventually became Sony, followed quickly with the first of many models in 1955, launching an industry that dominated Japanese business for generations.

It wasn't just portability that drew the public to the transistor radio. It was also the brand-new tunes playing over the radio: rock and roll.

15 MODEM

INVENTOR BRENT TOWNSHEND | 1996

DATA, THE LIFEBLOOD of electronic communications, needs a translator. One of the most important breakthroughs arrived in 1949, when the first modem converted U.S. Air Force radar data into sounds and squawked them over phone lines. A receiver then translated those terrible noises back into data.

Speeds multiplied rapidly during the 1980s, allowing the delivery of faxes at what then passed for lightning speed. (Now it would feel like an eternity.) By the mid-'90s internal PCI modems were a given. Transmission was slow until the late 1990s when Dr. Brent Townshend developed a version that hit 56 kbps, significantly faster than the previous 38.6 kbps speed—even if users had to put up with a nasty hissing sound that let them know they were about to be connected. Soon broadband Internet access became the norm, with USB modems bridging the gap from dial-up.

THERE'S NO RHYME OR REASON TO THE ENGLISH ALPHABET. IT WASN'T DESIGNED WITH THE MOST POPULAR LETTERS IN THE FRONT OR THE TALLEST LETTERS IN THE BACK. AND THAT WOULD BE JUST FINE . . . IF YOU WEREN'T TRYING TO TYPE IN THE 1800s.

Back then, writing machines jammed easily and were "full of caprices, full of defects—devilish ones," to quote Mark Twain. That's because a typewriter keystroke caused a mechanical action—each strike moved a bar up to hit the writing surface. If neighboring bars were struck in quick succession, the whole mechanism jammed.

In the 1870s Christopher L. Sholes studied letter-pair frequency (which letters are used together most often, such as *th*) and reorganized the letter–key layout. The resulting qwerty keyboard—named for the beginning of its upper left row—was first introduced on the Remington Standard 2 typewriter in 1874. It prevented type bars from crossing and, by 1910, became the standard model for typewriters, surviving to this day in the form of the (much quieter) computer keyboard.

RELATED

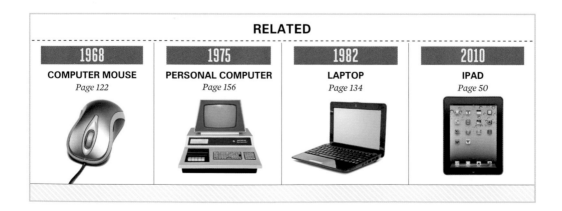

1968	1975	1982	2010
COMPUTER MOUSE	**PERSONAL COMPUTER**	**LAPTOP**	**IPAD**
Page 122	*Page 156*	*Page 134*	*Page 50*

MATCH

▶ In 577 A.D. the Chinese were packing pinewood sticks with sulfur. But it took nearly another thousand years for matches to make their way to Europe.

The next version was a combination of potassium chlorate, sulfur, sugar, and rubber. Lighting the match meant dipping its tip in a bottle of sulfuric acid. It was innovative—and very dangerous.

Striking became part of the process in 1826, when English chemist John Walker patented and sold "lucifer matches" with a modified formula that fired up when struck against a rough surface. Unfortunately, the lucifers were prone to sparking and emitted a nasty odor.

The addition of white phosphorus in 1830 solved the stink. The only problem: packs had to be kept in airtight containers, as they contained enough of the smoke-producing agent to kill the user and those around him.

Tinkering continued. **The first safety match was invented in 1898 by two French chemists, Henri Savene and Emile David Cahen.** But the use of white phosphorus wasn't completely eliminated until 1910, when President Taft asked the Diamond Match Company to release the patent for sesquisulfide phosphorus for the benefit of all mankind.

Since then, the match and its high-tech sequel, the lighter (see page 63), have transformed a wide range of activities, from cooking to recreational drug use. You can still buy them for less than a penny a piece or get free matchbooks at bars or restaurants, even if they don't allow smoking.

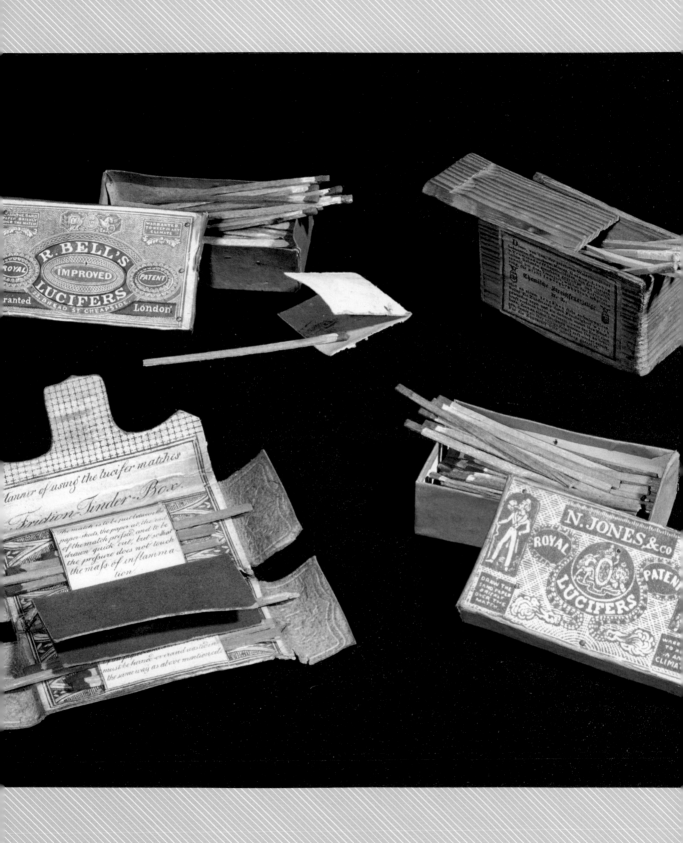

BICYCLE

INVENTOR KARL VON DRAIS | 1818

WITH NO PEDALS, CHAINS, OR BRAKES, the first bicycles required riders to push their feet on the ground for forward motion. First patented in 1818 and evolving steadily through the last half of the 1800s, the bicycle had a revolutionary impact not just on personal mobility, but also on the women's rights movement.

How? It helped bustle-wearing women change into "rational dress" much better suited to bicycle riding. "I think [bicycling] has done more to emancipate women than anything else," Susan B. Anthony said in 1896. "It gives women a feeling of freedom and self-reliance. I stand and rejoice every time I see a woman ride by."

The impact of bicycles didn't stop with the suffragettes.

◄

The larger front wheel of the Victorian penny-farthing enabled greater speeds and a smoother ride, but sitting high over the front axle often meant flying off the bicycle headfirst. On downhill journeys, riders often threw their legs over the handlebars for a safer, feet-first crash landing.

Bicycle manufacturers were pioneers in the development of mass production techniques, vertical integration, and even vehicle advertising. Try turning the page of a magazine from the turn of the 19th century without seeing an ad for bicycles. It also changed the way kids delivered newspapers.

And let's not forget another important contribution: Orville and Wilbur Wright ran a bicycle repair shop while figuring out how to fly over Kitty Hawk.

(11) DRY CELL BATTERY

INVENTOR GEORGES LECLANCHÉ | 1866

VOLTAIC PILE may sound like a kind of carpet, but the name actually refers to an early wet-cell battery invented by Alessandro Volta in 1800. His brine-soaked cardboard sandwich made with zinc and copper discs paved the way for decades of development in storing and accessing electricity.

The breakthrough in dry cell technology came when Frenchman Georges Leclanché found an ammonium chloride paste to sub for the liquids and figured out how to seal it effectively. In 1899, two years before Thomas Edison made his mark

with alkaline storage batteries, Waldmar Jungner invented the nickel-cadmium rechargeable battery.

The items we put batteries into have evolved dramatically over the years. But the dry cell battery is more or less the same. It still surrounds an electrode with a paste and houses it in a metal casing.

▶ The basic design of the incandescent bulb hasn't changed much since Thomas Edison developed it by gaslight and flicked the first switch in 1879.

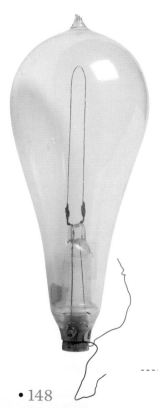

(Of course, Edison isn't the only person to have conceived of the idea, but he's the guy who made it practical for household use.) Stated simply, electrical current makes a filament glow white-hot. Thus converted to heat, some of this electric energy gets released as visible light. At some point—traditionally about 1,000 or so hours—the filament can't take any more, the bulb burns out, and you get another one out of the closet.

It's difficult to underestimate the impact of this invention. By turning night into day, it changed the way we work, play, travel, and celebrate holidays. And it made General Electric the only original Dow Jones index company still part of that club today.

"HOW MANY FILAMENTS DOES IT TAKE TO CHANGE A LIGHTBULB?"

Edison experimented with a number of materials before settling on a carbon filament in his 1879 lightbulb patent. That didn't stop other inventors from trying to make it better. In 1904 Hungarian inventor Sándor Just and Croatian inventor Franjo Hanaman patented a tungsten filament that burned longer and brighter light than carbon. When General Electric found a way to make filaments from sintered tungsten, it became the standard.

ALARM CLOCK

INVENTOR ANTOINE REDIER | 1847

THE ALARM CLOCK, ONE OF MODERN TIME'S MOST IMPORTANT—AND RESENTED—DEVICES, DATES BACK TO ANCIENT GREECE, WHERE PLATO SUPPOSEDLY HAD ONE.

Other stories tell of a Buddhist monk concocting a striking clock in the 10th and 11th centuries. In the mid-1800s French inventor Antoine Redier patented an adjustable mechanical alarm clock. About two decades later, the Seth Thomas Clock Company did the same with a small bedside alarm.

By the 1880s major U.S. clock companies were marketing them, and Germany quickly followed. They were in such demand that, when World War II-era factories went back to peacetime manufacturing, the alarm clock was one of the first products on the line. Why? Because workers had been missing their shifts due to, you guessed it, worn-out alarm clocks.

The ensuing decades brought alarm clocks that worked remotely, flashed lights, and woke sleepers to their favorite radio stations. In 1956, they even started allowing for a little **extra snooze time at the touch of a button.** Early snooze buttons offered 5- or 10-minute options, but 9 minutes became the standard in the '60s.

WHY 9 MINUTES? LET ME SLEEP ON THAT.

WHEN THOMAS EDISON UNVEILED the first device to record and play back sound in 1877, he predicted its primary use to be automated messages delivered over phones. Luckily, other uses revealed themselves as his embossed tin cans gave way to wax cylinders and vinyl discs.

Before they were staples in the home, though, phonographs were attractions along the lines of movie theaters. Customers at phonograph parlors in cities throughout the U.S. would, for a nickel, have the pleasure of hearing their personal selection played through a tube.

During early mass production of recorded music, machines could only produce 90 to 150 copies of a single performance. Musicians had to record the same song up to 50 times a day in order to keep up with demand.

Formats evolved, with 78-rpm, 45-rpm, and eventually 33⅓-rpm vinyl discs taking turns as the new standard. **Phonograph recordings were the centerpiece of the home's music system for much of the 20th century,** withstanding the inventions of both 8-track and cassette tapes. Then along came CDs, and the reign was over—although stalwart fans continue to collect and create vinyl recordings.

TELEPHONE

INVENTOR ALEXANDER GRAHAM BELL | 1912

▶ *"As a practical man, I did not quite believe it; as a theoretical man, I saw a speaking telephone by which we could have the means of transmitting speech and reproducing it in distant places. But it really seemed too good to be true…"*

—Alexander Graham Bell,
***Popular Mechanics*, 1912**

Few gadgets have as familiar an eureka moment as that of the telephone. And none besides the telephone evolved into something people now carry with them every waking moment.

The first version available for commercial purchase looked more like a box camera than today's phone, with the same opening serving as a transmitter and receiver. It was as alien to the twisty-corded, wall-mounted unit that late-20th-century chatters used as those phones are to cordless phones and mobiles of users today.

The ability to communicate instantaneously and privately with a person on the other side of the globe not only improved commerce (from pizza delivery to Fortune 500 business) and made it easier to reach out and touch family, it also changed our sense of the individual. Whatever the shape or size, the telephone gave us voice beyond our offices and living rooms.

▶ **Though available to the public since 1963, push-button phones were mostly used by businesses. Until the 1980s the majority of home phone calls were made on a rotary dial.**

WHEN WILLIS HAVILAND CARRIER AND HIS SCANTILY CLAD FEMALE ASSISTANTS DEMONSTRATED THE FIRST ROOM AIR CONDITIONER FROM AN IGLOO AT THE 1939 WORLD'S FAIR, THE WORLD SUDDENLY SEEMED A WHOLE LOT COOLER.

Prior to the development of the window unit, if you wanted air conditioning in an older home, you had to go through massive and costly renovation. Now winter's chill was available coast to coast, all year long. But before air conditioning could really catch on, World World II began and Carrier turned its production to military use. The portable air conditioner didn't have its day in the sun, so to speak, until the post-war economic boom of the 1950s.

YEAR-ROUND COMFORT
To counter a crop of cheap imitators, 11 companies met in 1935 to officially define "air conditioning." They declared that true air conditioners should cool, dehumidify, and circulate air in summer, while heating, humidifying, and circulating air in the winter.

STEVE JOBS
Factor in the work of Steve Jobs, whose simple and elegant Apple devices pushed design to the forefront, and you have the biggest revolution in American consumerism since the automobile.

PERSONAL COMPUTER

INVENTOR VARIOUS | **1975**

BEFORE THERE WERE COMPUTERS, there were "computers"—World War II–era women performing ballistic calculations on the University of Pennsylvania's Differential Analyzer. On the very same campus, scientists were developing the first general-purpose electric computer, ENIAC, which would become the main computing machine for the U.S. Army.

That computer weighed 30 tons and took up the better part of a 30 x 50–foot room—even though it didn't have the computer capability of an average laptop today. The punch cards that carried the data were innovative, but bulky and unwieldy.

What changed? Microchips, for one. In 1996 students at Penn were able to pack the computing power of ENIAC onto a single microchip.

Computers became personal with the 1975 debut of the Altair 8800, a hobbyist kit with no defined purpose. Main-stream minds were not exactly blown away, but the geeks saw the binary on the wall. Within months Bill Gates released a programming language for the Altair, and by 1977 the Apple II revealed the personal computer's true potential: It shipped with the video game *Breakout*.

Until 1981 IBM had mostly been focused on large-scale mainframe computing for businesses, but with the rise in the late 1970s of "microcomputers" such as the Apple II and Tandy TRS-80, IBM set out to make a desktop computer system designed for an individual. While the PC may not have been the first true "personal computer," it became synonymous with the technology. Through IBM's partnership with the fledgling Microsoft company, the PC set in motion a movement toward operating-system standardization (and gave Microsoft a near monopoly in the OS business for years).

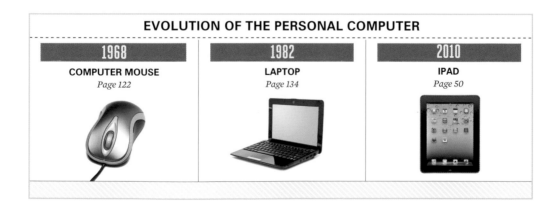

EVOLUTION OF THE PERSONAL COMPUTER

1968	1982	2010
COMPUTER MOUSE	**LAPTOP**	**IPAD**
Page 122	*Page 134*	*Page 50*

THE PREMISE OF THE SYRINGE—INJECTING DIRECTLY INTO THE BLOODSTREAM—ISN'T ALL THAT NEW. YOU CAN JUST LOOK AT THE HONEY BEE TO SEE THE BASIC TECHNOLOGY.

Experiments with piston syringes date back to the ancient Greeks and Romans. But it wasn't until the mid-1800s that French physician Charles Pravaz and Scotsman Alexander Woods independently developed working syringes. The first disposable sterile syringe, created as a way to vaccinate animals, was patented in 1956. The first plastic ones were introduced in 1974.

All good? Not exactly. Woods's wife, who died of a self-administered morphine overdose, might be the first case of death by syringe. And the deaths of recreational drug abusers are untallied. Then there are the solid-waste problems and high hospital expenses that come with single-injection hypodermics.

Still, you can't dispute the positive **impact of the syringe, which has saved millions from polio, tuberculosis, small pox, and more.**

○ NEEDLE SHARP

YOUR DISTANT ANCESTORS DIDN'T HAVE TO WORRY ABOUT TRYPANOPHBIA (FEAR OF NEEDLES) BUT IT'S A GOOD THING THEY EXIST TODAY.

2011

221
Needle exchange programs active in the United States

More than
36
MILLION
Syringes were distributed annually

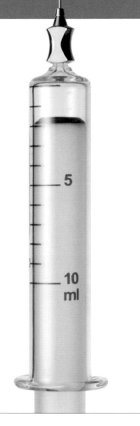

COST PER NEEDLE
US $0.97

COST OF ONE DAY'S WORTH OF THE HIV TREATMENT MEDICATION

US $36

Needles range from

7 gauge
(the largest to
33 (the smallest)
on the Stubs scale

21-gauge needles
are most commonly used for drawing blood for testing purposes

16- or 17-gauge needles
are most commonly used for blood donation

10%

of the adult population has a phobia of needles

"TELEVISION IS READY FOR THE HOME!"

trumpeted scientist Dr. Vladimir K. Zworykin in a 1933 *Popular Mechanics* article. He jumped the gun a bit—but only by a few decades.

Conceptualized in 1877 by Thomas Edison and Alexander Graham Bell and patented by inventor Philo T. Farnsworth in 1930, television as a concept was fairly well-known by the 1939 World's Fair. But the masses didn't started bringing them into their living rooms until World War II ended and crucial materials became available again. Regular network television began broadcasting, TV trays and stars were born, and radio's days as the top medium were over and out.

RELATED

1950

REMOTE CONTROL
Page 136

POLITICS CHANGED, TOO.

"We wouldn't have had a prayer without that gadget," John F. Kennedy said of the tube, since most of the 75 million Americans who watched the debates between Kennedy and Nixon during the 1960 election gave the nod to the senator from Massachusetts, while those who listened on radio thought Nixon had won.

TELEVISION'S NUMBERS GAME

IN THE POST-WAR DECADES, INCREASINGLY SOPHISTICATED GADGETRY GIVES THE U.S. HOUSEHOLD EVER-SHARPER VIEWS THROUGH ITS WINDOW ON THE WORLD.

1950s

HOUSEHOLDS WITH TVS

1950
9 percent
(3.8 million)

1959
85.9 percent
(43.9 million)

4

Number of major national networks—NBC, ABC, CBS, and the DuMont Television Network—that provide broadcast content to local affiliates

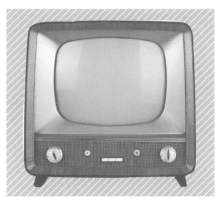

$150

Price of Zenith's 17-inch Trend-Setter TV in **1955**

$240

Price of Magnavox's 21-inch Telerama 21 in **1955**

1960s

$349

Price of Admiral's 23-inch Stewart color set in **1968**

$28 MILLION

Fee CBS pays in **1964** for broadcast rights to the NFL. The network introduced instant-replay technology the previous year

CONTINUED ➜

1960s

1970s

HOUSEHOLDS WITH TVs IN 1969

95 percent
(58.2 million)

NBC

Becomes the first U.S. network to complete conversation of all new programs to color **(1966)**

COLOR TV SETS

Outnumber black-and-white sets in U.S. homes for the first time **(1972)**

By **1978**, 78 percent of U.S. households with TVs have color sets

TELSTAR 1

This communications satellite, launched in 1962, transmits events in real time from all corners of the world

SEPTEMBER 30, 1975: HBO becomes the first network to broadcast signals continuously delivered by satellite for the **"Thrilla in Manila" Ali vs. Frazier boxing match**

MONDAY NIGHT FOOTBALL
debuts on ABC in 1970

PBS

Is the first network to deliver all its programming via satellite instead of landlines **(1978)**

1980s

HOMES WIRED FOR CABLE

1980
20 percent

1985
43 percent

98%

U.S. households that have television, half of them with multiple sets. In **1980**, 83 percent of TVs are color sets

In **1985** VCRs are in 14 percent of U.S. homes

1990–

March 2007: Digital tuners are required for new TVs that receive signals over the air

June 2009: TV stations cease analog broadcasting, completing the transition to digital broadcasting

AVERAGE COST FOR AN HDTV

2004	2009	2011
$1500 »	**$731** »	**$609**
6 percent of homes have one HD set	**50 percent** of homes have one HD set	**72 percent** of homes have one HD set

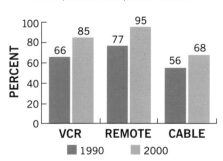

65 percent of homes have at least two TVs in **1990**

PERCENTAGE OF HOMES WITH VCRS, REMOTES, AND CABLE

VCR: 66 (1990), 85 (2000)
REMOTE: 77 (1990), 95 (2000)
CABLE: 56 (1990), 68 (2000)

PERCENT — ■ 1990 ■ 2000

NUMEBER OF HOUSEHOLDS WITH DVD PLAYERS

6.7%
1999

81%
2006

Average net paid circulation for September exceeded
Daily — 1,800,000
Sunday - 3,150,000

DAILY NEWS

Copyright 1938 by News Syndicate Co., Inc. Reg. U. S. Pat. Off. NEW YORK'S
PICTURE NEWSPAPER
Entered as 2nd class matter, Post Office, New York, N. Y.

 FINAL

Vol. 20. No. 109   New York, Monday, October 31, 1938★   48 Pages   2 Cents IN CITY LIMITS | 3 CENTS Elsewhere

FAKE RADIO 'WAR' STIRS TERROR THROUGH U.S.

—Story on Page 2

"War" Victim

Caroline Cantlon, WPA actress, listening to this radio in West 49th St., heard announcement of "smoke in Times Square." Running to street, she fell, broke her arm.

(NEWS foto)

"I Didn't Know".

(By Associated Press)

Orson Welles, after broadcast expresses amazement at public reaction. He adapted H. G. Wells' "War of the Worlds" for radio and played principal role. Left: a machine conceived for another H. G. Wells story. Dramatic description of landing of weird "machine from Mars" started last night's panic.

—Story on page 2.

2 RADIO

INVENTOR GUGLIELMO MARCONI | 1896

IN A TELEVISION-AND-COMPUTER-DRIVEN WORLD, IT'S EASY TO OVERLOOK THE REVOLUTION WROUGHT BY THE RADIO. BUT TRY TO IMAGINE A SILENT CAR RIDE, OR SITTING ALONE AT HOME AS A STORM BREWS WITHOUT KNOWING ITS POTENTIAL POWER.

Patented in England in 1896 as "wireless telegraphy" by Guglielmo Marconi—who based his work on technology developed by Nikola Tesla—the first radios were used primarily for maritime communication. By 1938, though, they were in 80 percent of U.S. homes. Radio was the first electronic form of mass communication, giving broadcasters a bigger megaphone than had ever been seen—or heard—on Earth.

THE WAR OF THE WORLDS

The day before Halloween 1938, Orson Welles's *Mercury Theatre on the Air* presented what sounded like a news broadcast of a Martian invasion but was actually an adaptation of H.G. Wells's *The War of the Worlds*. Police switchboards jammed. Drivers fled cities. Doctors volunteered to treat the injured. And panicked or not, everyone saw the power of mass media.

MOBILE/SMARTPHONE

INVENTOR AMOS E. JOEL, JR. | 1971

ITS ORIGINS TRACE BACK TO FINLAND AND JAPAN IN THE '70S AND SHIP-TO-SHORE EXPERIMENTS AS FAR BACK AS THE 1940S.

But mobile phones weren't on the road to becoming the most widely used gadgets in the world until Amos E. Joel, Jr., a Bell Labs engineer, invented a system that allowed for uninterrupted service between cell areas.

That was in 1971. And with the prospect of calling without disconnects, the technology began evolving exponentially. Eleven years after Joel's jump, the FCC approved AT&T's proposal for an Advanced Mobile Phone System and allocated frequencies.

Technology leapt forward in 1983 with the Motorola DynaTAC 8000x, the first truly portable cellphone. Once a gadget of the rich and powerful (one iconic cinematic use: Michael Douglas in Wall Street), cellphones quickly morphed into a product category second to none. It took 20 years to sell the first billion units. The second billion sold in four and the third billion in only two. By the end of 2010 the subscription rate stood at 5 billion, or 75 percent of all people on earth.

Nowadays, "phone" is hardly the right word. Enabled with wireless, GPS, and multimedia, today's smartphone—more of a pocket PC—facilitates instantaneous personal connections that make house-to-house phone conversations seem like cave paintings. And those phones are disappearing: More than one in four U.S. households have cellphones but no landlines.

Now people of developing nations—even those without an electrical grid—can tap into the world's commerce and culture via smartphone. Or just text a friend.

BRAINS VS. BRAWN

The 10-inch Motorola DynaTAC tipped the scales at 2.5 pounds and needed to be recharged after an hour-long phone conversation. The 4.9-inch iPhone 5S may seem bulkier than the clamshell you had back in the day, but it's still just under 4 ounces. And say what you will about battery life—it still provides an estimated 10 hours of 3G talk, 10 hours of Wi-Fi, and 40 hours of audio playback.

PHOTO CREDITS

105 inset, 113 top left, 118 inset, 118 bottom, 120 top left, 123 bottom left, 126 top, 134 bottom left, 135 bottom, 137 left, 138 bottom, 143, 150, 152, 159 right, 161 (both), 162 (all photos), 163 middle, 163 right, 165 bottom right; Edward Westmacott: 13 top; Catherine Lane: 18 top left, 107 bottom right; Matt Naylor: 18 top right, 107 top; Daniel Stein: 20; AKS Photo: 21 top, 29 bottom left; Coprid: 31 top; Burwell Photography: 39 top; Kiyoshino: 46 bottom middle, 64 bottom middle, 94 top, 112 bottom left; Rinek: 50 bottom (second from right), 123 bottom (second from left), 134 bottom (second from right), 142 bottom (second from left); Floortje: 55 bottom right; Hulton Archive: 60 bottom; Scott Kochsiek: 69 bottom; Don Nichols: 76 top right, 135 middle right; Studio 58: 125 bottom; Magnet Creative: 159 left; Karen Mower: 167 right
The Kobal Collection of Art Resource, NY: 126 bottom
Martin Laksman: 4
Library of Congress: 51, 88–89, 146
Mary Evans Picture Library: 37, 78; National Magazines: 61; Retrograph Collection: 102
NASA: 53 left
National Archives: 100
National Geographic Creative: Bates Littlehales: 38; Gordon Wiltsie: 58
Michael Schmelling: 71 bottom, 104 inset
Shutterstock: 8 top left, 27, 36 bottom, 41 (both), 46 top, 48, 52, 56 left, 64 bottom right, 69 top, 70 bottom, 77 bottom, 86, 94 bottom middle, 103 top, 107 bottom left, 110, 112 middle bottom, 113 top right, 114 bottom left, 119, 120 bottom, 125 top left, 138 top right, 167 left; Gladkova Svetlana: 10 top; Winai Tepsuttinun: 13 bottom, 14 bottom right; Ton Lammerts: 14 left; Africa Studio: 15 top; Richard Peterson: 28; Carolina K. Smith MD: 29 top; Aaron Amat: 31 bottom; Planner: 34 left; Raymond Kasprzak: 36 top; Biehler Michael: 42; Bob Orsillo: 43 right; Claudio Divizia: 46 bottom left, 64 bottom left, 94 bottom left; Paul Paladin: 46

bottom right, 94 bottom right, 112 bottom right; Kinetic Imagery: 50 bottom (second from left), 134 bottom (second from left), 142 bottom left; Tatiana Popova: 50 bottom right, 123 bottom (second from right), 142 bottom (second from right); Quang Ho: 53 right; Aeromass: 55 top; Lusoimages: 62 bottom left, 76 top left, 135 top; H. Brauer: 62 bottom right, 76 bottom, 135 middle left; Maxx-Studio: 66; Nikitin Victor: 67 inset; Petr Malyshev: 79 top, 84 right; MalDix: 80 top; Evgeny Karandaev: 96; Berci: 105 bottom; Yuri Tuchkov: 111 top left, 114 bottom right; Przemyslaw Ceynowa: 115 bottom; Oleksiy Mark: 117 top; Alaettin Yildirim: 117 bottom; Nikodem Nijaki: 133; Carlos E. Santa Maria: 137 right; Vasiliy Ganzha: 140; Jim Pruitt: 141; Everett Collection: 155
Stockfood: Rua Castilho: 12
Studio D: 75 top, 108 left; Philip Friedman: 111 top right, 114 top, 125 top right
SuperStock: 154; Science and Society: 18 bottom, 134 top, 145

FRONT COVER (clockwise from top left): iStock, Courtesy of iRobot, iStock, Jeremy Pembrey/Alamy (Polaroid and Polaroid & Pixel are trademarks of PLR IP Holdings, LLC, used with permission), Courtesy of Victorinox Swiss Army, Shutterstock
INSIDE FRONT FLAP: iStock
SPINE: iStock
BACK COVER (clockwise from top left): Maxx-Studio/Shutterstock, iStock, Shutterstock, Nikodem Nijaki/Shutterstock, iStock, iStock

INDEX